AS/A2 GEOGRAPHY

CONTEMPORARY CASE STUDIES

The Energy Issue

David Holmes and Sue Warn

PHILIP ALLAN
UPDATES

Philip Allan Updates, an imprint of Hodder Education, an Hachette UK company, Market Place, Deddington, Oxfordshire OX15 0SE

Orders
Bookpoint Ltd, 130 Milton Park, Abingdon, Oxfordshire, OX14 4SB
tel: 01235 827827
fax: 01235 400401
e-mail: education@bookpoint.co.uk

Lines are open 9.00 a.m.–5.00 p.m., Monday to Saturday, with a 24-hour message answering service. You can also order through the Philip Allan Updates website: www.philipallan.co.uk

© David Holmes and Sue Warn 2011

ISBN 978-1-4441-1983-1

First printed 2011
Impression number 5 4 3 2
Year 2-16 2015 2014 2013 2012 2011

Front cover photograph ©BP plc

Printed in Italy

Hachette UK's policy is to use papers that are natural, renewable and recyclable products and made from wood grown in sustainable forests. The logging and manufacturing processes are expected to conform to the environmental regulations of the country of origin.

P01869

Contents

Part 5: The nuclear question

Part 6: Renewable opportunities

Part 7: The future energy challenge

Part 8: Examination advice

Index

Introduction

> Historians now see the turn of the 19th century as the dawn of the Industrial Revolution.
> I hope they will see the turn of the 21st century as the dawn of the energy revolution.
>
> *Rob Routs, Shell, June 2007*

There can be no doubt that energy can both enhance and threaten what we value most — health, community and the environment, the future of our children, and the planet itself.

For advanced economies, reliable energy supports the technologies and services that enrich and extend life. For example, it powers computers, transportation, communications and cutting-edge medical equipment.

For developing and emerging nations, establishing reliable and affordable supplies of energy accelerates changes that improve quality of life. Providing access to energy means expanding industry, modernising agriculture, increasing trade and improving transportation. These are the building blocks of economic growth, which create the jobs that help people escape poverty and create better lives for themselves and their children.

We need to consider some of the complex issues relating to types of energy and their unequal distribution of supply round the globe. This spatial pattern of energy supply and consumption in itself is most interesting, not least because it links strongly with development, power and geopolitics. A variety of environmental, economic, political and technological factors have led to rapid changes in the geography of energy.

We also need to consider the very real issues of our existing energy mix, such as an over-reliance on fossil fuels and potential electricity 'crunches'. People around the world have very different ideas on how we deal with the future of energy supply together with changing demands for gas and oil. What are our responsibilities in considering nuclear energy, renewables and energy conservation?

These factors combine to make energy a highly important, emotive and controversial topic for geographical study.

About this book

This book explores the geography of energy under several different headings. It starts with an introduction to types of energy and its supply, then considers the oil, gas and nuclear options. The book concludes with a discussion of renewable energy and future energy options, including energy conservation. Each part of the book contains several up-to-date case studies, which support the main ideas in each topic.

Parts 1 and 2 look at the different types of energy (renewable, non-renewable and recyclable) and consider the range and types of energy sources as well as their spatial patterns of distribution. The mismatch between reserves and consumption (see Figure 1) is discussed, with a focus on the geopolitics of conflict and cooperation in certain parts of the world. Overarching case studies explore some of the key issues of energy supply and conflict, including the 'gas wars' in Europe and energy insecurity in the USA.

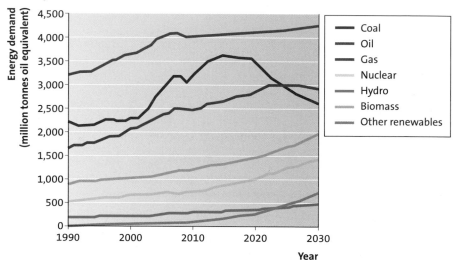

Figure 1
World energy demand

Part 3 deals with the issues of existing energy supply and consumption, considering externalities as well as social and economic impacts. Environmental concerns are also examined in Part 3, with a particular focus on the link between over-dependence on fossil fuels (especially coal) and carbon footprints.

Parts 4 and 5 focus on the current and future issues surrounding gas, oil and nuclear power. The controversial theory of peak oil is discussed and evaluated, together with the main oil players now and in the future. Three case studies look at exploration for oil and gas at the margins of what is either technically or socially feasible. The ongoing and contentious debate surrounding nuclear power is re-examined, with a focus on its green credentials and the economics involved. Is the advance of nuclear power and associated knowledge a development pathway that is particularly attractive for emerging economies? What are the possible conflicts associated with such expansion?

Parts 6 and 7 look at the potential of renewable technologies along with future energy challenges. Do renewables really offer a universal panacea or are they a non-starter? Part 6 tries to take a pragmatic point of view in the light of conflicting evidence about the utility of different renewable energy sources, including hydroelectric, wind and biofuels. Case studies include a discussion of the UK's

Contemporary Case Studies

controversial Severn Barrage and an assessment of geothermal energy in Iceland. Part 7 offers some sobering thoughts relating to future demand predictions (again, see Figure 1), the need for energy conservation and the much-hyped energy 'gap'. Can technology really secure the future or is there a need for greater energy conservation?

Part 8 provides guidance on:
- answering AS and A2 exam questions
- developing essay-writing skills
- researching energy topics

Key terms

BRICs: the emerging economies of Brazil, Russia, India and China.

Carbon-neutral: having a net zero carbon footprint, i.e. achieving net zero carbon emissions by balancing a measured amount of carbon released with an equivalent amount sequestered or offset.

Carbon pricing: an administrative approach imposing a cost on the emission of greenhouse gases that cause global warming. Paying a price for carbon released is a way of motivating countries, businesses and individuals to reduce emissions.

Energy mix: the range of energy sources used.

Energy security: the uninterrupted access to reliable sources of supply at an affordable price, with exploitation not having an undue effect on the environment.

Externalities: an economics term to describe the 'spill-over' effects arising from the production and/or consumption of goods.

Flow resources: continuous sources of energy such as tidal, wind and solar.

Fossil fuels: carbon-based fuels formed in previous geological times from buried organic material. They include peat, coal, oil and gas.

GDP: gross domestic product — a measure of the wealth of a region or country through its overall economic output.

Geo-engineering: usually taken to mean proposals to deliberately manipulate the Earth's climate to counteract the effects of global warming from greenhouse gas emissions.

Grid: the infrastructure of electricity or other energy transmission lines (or pipes) that connect power stations to the end user, i.e. homes, offices, factories etc.

HDI: human development index — a quantitative compound index used to rank countries in terms of development.

Liberalised electricity markets: where industries and private households are in theory able to freely choose their energy supplier. This should introduce competition and reduce prices.

LNG: liquid natural gas.

Nuclear fuel cycle: the processing of uranium ore through various stages to enrich it before it is used to generate electricity.

Off-grid: also known as **off-the-grid (OTG)**, this refers to living in a self-sufficient manner without reliance on one or more services (in the context of this book, having no electricity).

OFGEM: the Office of Gas and Electricity Markets. It regulates gas and energy markets in the UK.

Peak oil: the point at which the rate of consumption of oil reserves is not matched by new supplies coming on stream.

Player: an individual, group or organisation that has a vested interest in a development or proposal. Often there is conflict between different groups of players.

Primary energy: energy produced by using raw materials directly, such as coal for heating.

Recyclable energy: where fuel that has been used once can be recycled to be used again. At present only nuclear power is classified as recyclable through complex reprocessing.

Resource: any part of the environment that can be exploited by humans to meet their needs.

Secondary energy: a form of energy produced from primary energy sources, e.g. electricity or petroleum.

Stock resources: finite and non-renewable sources of energy such as fossil fuels.

toe: tonne of oil equivalent — a universal measurement of energy consumption and usage.

Websites and further reading

www.ukerc.ac.uk/support/tiki-index.php — The UK Energy Research Centre.

www.energysavingtrust.org.uk — The EST provides ideas on how householders can reduce their energy footprints, for example advice on buying energy-saving appliances.

www.nea.org.uk — National Energy Action develops and promotes energy efficiency services to tackle the heating and insulation problems of low-income households.

http://geology.com/research/barnett-shale-gas.shtml — An article about the new shale gas deposits in Texas.

www.shell.com — Shell's global homepage.

www.gazprom.com — A profile of the Russian energy TNC.

www.opec.org/opec_web/en — OPEC's global home page is a useful source of World Oil Outlooks.

www.gecforum.org — The Gas Exporting Countries Forum.

www.internationalrivers.org/en/about-international-rivers — International Rivers Website holds information and reports about possible problems of mega dams.

www.russiaprofile.org — A hunt through this extensive website reveals interesting energy facts about this emerging superpower.

www.bbc.co.uk — A huge website that has a host of articles relating to energy. Use the search facility to find them.

www.economist.com — *The Economist* has a wealth of energy-related articles. Use the search facility to access some of them. Note that some require subscription.

www.newscientist.com — *New Scientist* is a leading science magazine written in an accessible format. It has a wealth of energy-related articles. Use the search facility to access some of them. Note that some require subscription.

www.energynow.com — Lots of useful environmental energy information.

www.gwec.net — The global forum for the wind energy sector, uniting the wind industry and its representative associations.

www.worldbank.org — The World Bank's huge website includes lots of useful energy reports.

www.odi.org.uk — Website of the Overseas Development Institute.

www.decc.gov.uk — Website of the UK Department for Energy and Climate Change.

www.iaea.org — Website for the International Atomic Energy Agency.

www.ren21.net/ — Website for the Renewable Energy Policy Network for the Twenty-First Century.

www.carbontrust.co.uk — This provides specialist support to businesses and the public sector to help cut carbon emissions, save energy and commercialise low-carbon technologies. The emerging technologies section is especially useful.

Useful books

Cravens, G. (2008) *Power to Save the World: The Truth about Nuclear Energy*, Vintage.

Garrington, S. (2011) *Energy: The Burning Questions*, Geographical Association.

Nikiforuk, A. (2010) *Tar Sands: Dirty Oil and the Future of a Continent*, Greystone Books.

Price, N. (2007) *The Energy Crisis: How Do We Fuel Our Future?* Pocket Issue.

Wengenmayr, R. (ed.) (2008) *Renewable Energy: Sustainable Energy Concepts for the Future*, Wiley.

Types of energy

The importance of energy

Energy is central to the functioning and development of human societies. In the nineteenth century coal powered the Industrial Revolution. In the twentieth century oil and gas powered radical improvements in levels of affluence and productivity of millions of people throughout the world. And in the twenty-first century, energy has become so vital to our lives that there is a growing realisation that energy systems will need to be changed radically if they are to supply our burgeoning energy needs on a long-term and equitable basis. We must also pay due regard to minimising environmental damage.

The standard scientific definition of energy is that it is 'the capacity to work'. The word **energy** is often incorrectly used synonymously with **power**, which is strictly 'the rate of doing work', i.e. the rate at which energy is converted from one form to another, or is transmitted from one place to another.

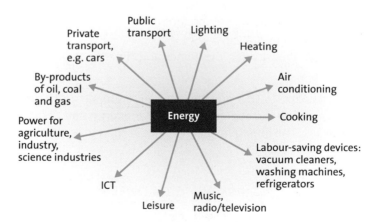

Figure 1.1
Energy: the vital resource

Figure 1.1 shows all the ways that energy has become vital to modern living and essential to human wellbeing. You can see this importance in developed countries: short term with the panics from tanker drivers' strikes or winter power cuts, and long term in countries such as South Africa, the USA and the UK, where future energy security is threatened. The prospect of the lights going out because of an energy gap caused by poor planning is very real.

In the developing world energy poverty is a major concern as it presents a huge brake on development and reinforces the development gap. Some 1.5 billion people, largely concentrated in sub-Saharan Africa and south Asia, exist without electricity (see Figure 1.2).

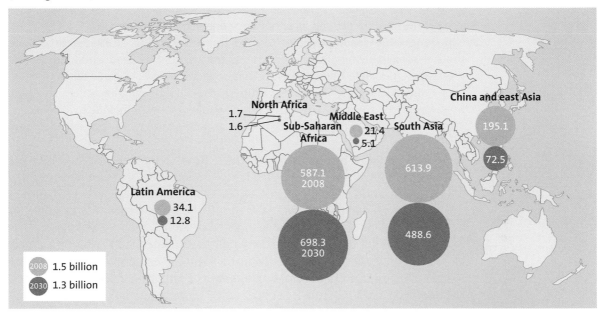

As Figure 1.2 shows, some improvements are expected by 2030, except in sub-Saharan Africa, where energy developers have little incentive to build power plants or connect remote areas to the grid if the people are too poor to pay. As you will see on pages 79–92 sustainable energy projects featuring micro HEP, solar and wind power have become a key strand in aid programmes often financed or carried out by specialist energy NGOs.

Figure 1.2
World population without access to electricity (millions)

Africa/Fotolia

Figure 1.3
Many people in rural areas of sub-Saharan Africa rely on fuelwood for energy

Energy classification

Resources

Energy **resources** are classed as either non-renewable or renewable.

■ **Non-renewable resources** (also termed finite, capital or stock resources) are those that have been built up over time, so they cannot be used without depleting their stock. Because their rate of formation is very slow — geologically formed over millions of years — they are **finite**. If they are heavily exploited, this puts pressure on the source of supply. Non-renewable resources include hydrocarbons (the **fossil fuels** — coal, natural gas and oil) and uranium ore used to generate nuclear power.

■ **Renewable energy resources** (also termed **flow** or income resources) have a 'natural rate of availability'. They yield a continuous flow that can be consumed in any given period of time, without endangering future usage, provided current use does not exceed net renewal rates during the same period. For example, as a result of a series of climate change-induced droughts, the HEP supplies of New Zealand are under threat because demand has remained static while the supply of water to drive the turbines has dwindled. Renewable energy resources include solar power, wind, HEP, wave and tidal power, geothermal energy and biomass sources.

Renewable resources can be further divided into two sub groups:

■ **Critical** or **recyclable resources** include biomass energy from forests, plants or animal waste, and nuclear power produced by recycling spent rods of uranium from existing nuclear power stations. These resources require prudent management to ensure sustainable use. If they are mismanaged, renewability will not be automatic and depletion will result. **Sustainable** management involves a carefully controlled system to ensure that the current level of exploitation does not compromise the ability of future generations to meet their needs (the Brundtland principle). Recycled energy development (RED) is increasing the range of options, for example harnessing waste heat from furnaces to produce electricity.

■ **Non-critical** or **everlasting** resources such as tides, waves, wind, running water and sunshine are fully renewable. Flows are continuous, but in some cases they can fluctuate, for example wind and solar power.

Figure 1.4 summarises the way in which energy resources can be classified.

Reserves

A **reserve** is the proportion of a resource that can be exploited under current economic conditions, and with available technology. It is a vital concept when reviewing the global availability of energy resources for the future.

Again, a two-fold classification is used:

■ **Recoverable reserves** are the amount of an energy resource likely to be extracted for commercial use, i.e. **proved** resources that are economically viable with current levels of technology.

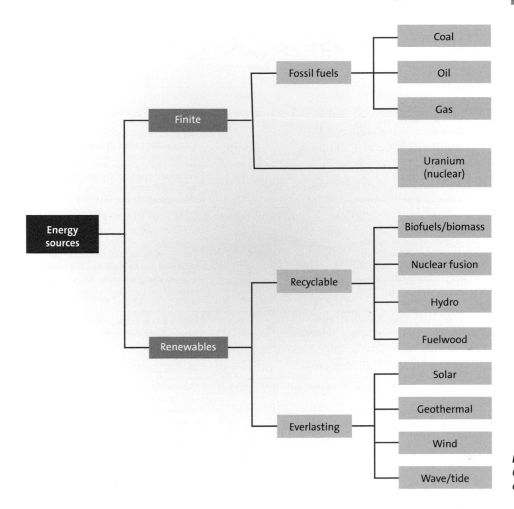

Figure 1.4
Classification of energy resources

- **Speculative reserves** are deposits where geological conditions parallel to existing operations suggest a likelihood of occurrence, but because of high technological costs they are **currently** not economically viable, or have not yet been explored. This makes it difficult to provide accurate predictions, for example for future oil and natural gas supplies. It is important that when you are researching this topic you always try to obtain up-to-date statistics.

Primary and secondary energy

Primary energy resources are raw materials, such as coal, natural gas, wood, sunlight, wind and waves, that are used in their natural form to produce power.

Secondary energy resources involve converting a primary energy source into a new form, such as crude oil into petroleum or coal into electricity.

It is important to look closely at any data tables and graphs to check whether they refer to primary energy production or consumption or to electricity. It is also vital that you have a basic knowledge of the units used — see page 116.

Table 1.1 (on p. 14) summarises the characteristics of the main energy resources.

Table 1.1 *Characteristics of the main energy resources*

Primary energy resource	Description	Classification	% of global energy supply, 2008	Key concerns/issues
Coal	A combustible, sedimentary rock formed of converted residual plant matter and solidified below overlying rock strata. There are several types of coal: hard/ bituminous coal, brown coal (lignite) and peat (strictly speaking, a precursor to coal).	Non-renewable	25%	Use releases large amounts of carbon dioxide and other pollutants, contributing to climate change and atmospheric pollution. Carbon capture technology for removing carbon dioxide from atmosphere unproven and complex.
Natural gas	A methane-rich gas found underground. It may also contain water vapour, sulphur compounds and other non-hydrogen gases such as carbon dioxide, nitrogen or helium.	Non-renewable	23%	Costs and security of supply, especially for countries that are largely importers. Releases carbon dioxide on use.
Nuclear fission	The division of a heavy nucleus into two parts, usually accompanied by the emission of neutrons (neutrally charged particles inside the nucleus), gamma radiation (high-energy radiation) and energy release. This energy is converted into heat that raises steam to drive turbines and generate electricity.	Non-renewable (may be recyclable with fuel reprocessing)	6%	Possible health risks associated with power plants and accidents such as Chernobyl. Disposal of radioactive material raises safety issues and there are unknown long-term risks. Amount of raw material left globally unknown.
Crude oil (petroleum)	A naturally occurring mineral oil consisting of many types of hydrocarbons. Crude oil may include small amounts of non-hydrocarbons. Also includes tar sands and oil shale.	Non-renewable	37%	Concerns that global supplies may have reached their peak, security of supply, geopolitical tensions and lack of alternatives, especially for transport. Releases carbon dioxide when burnt.
Solar	Energy directly harnessed from solar radiation, as distinct from wind, water and biomass energies indirectly driven by the sun. Solar radiation is absorbed by a collector and converted to heat energy, or into electricity by photovoltaic cells.	Renewable	0.5%	Distribution and availability varies spatially and temporally. Photovoltaic technology still expensive compared with fossil fuels.
Ocean	Energy harnessed by using either the physical characteristics of oceans (tidal movement, wave motion, thermal gradients, ocean currents) or their chemical characteristics (saline gradients).	Renewable	>0.1%	Only certain locations are suitable for offshore tidal generation. Technology for large-scale generation is unproven. Ocean sources have low energy densities, and so large numbers of devices are needed to harness this energy.
Wind	Directly related to solar activity, which causes differences in atmospheric pressure and temperature (and to the Earth's rotation and gravity). Modern wind turbines range from 600 kW to 5 MW of rated power.	Renewable	0.5%	Only certain locations have enough wind to be viable. Wind energy is variable power, so it is difficult to manage power supply through a grid system without some back-up.
Biomass	Organic, non-fossil material of biological origin. Although the different forms of energy from biomass are considered renewable, their rates of renewability differ. Wood is an example of a biomass energy source.	Recyclable	4%	Relatively low energy densities mean limited potential for large-scale electricity generation. Biomass acts as a carbon sink, so combustion releases stored carbon dioxide.
Hydrological	Energy harnessed from the movement of water through rivers, lakes and dams (owing to gravity). A 'head' of water is stored and then released to drive turbines and generate electricity. Hydroelectric systems can range in capacity from thousands of megawatts to small micro-hydro schemes.	Recyclable and renewable	3%	Large-scale systems are costly to build. Dam-building also has social, political and environmental impacts. Smaller micro-hydro plants may not be economically viable, but are vital for remote rural areas.
Geothermal	Comes from rocks within the Earth and can be tapped: (1) as hot water or steam, (2) as hot, dry rock energy and (3) by means of conduction. The first two are used to generate electricity while the third is used to heat water, buildings and greenhouses.	Renewable	0.2%	Geothermal heat in the outer 10 km of the Earth's crust is too diffuse to be exploitable worldwide. Availability is limited to a few locations such as Iceland and the Philippines.

Patterns of global availability

For a variety of reasons energy resources are distributed unevenly, with certain regions of the world having an energy surplus, and others an energy deficit. Inevitably this is exacerbated by an uneven demand (see Part 3) as consumption of energy can be closely linked to level of development. Indeed, energy use is often selected as a useful measure of economic development when creating country profiles.

Table 1.2 summarises the factors that contribute towards the unevenness of energy supplies. The underlying reasons for global variations in supply relate to physical factors, but availability can be very heavily influenced by economic and political factors.

Table 1.2 *Factors affecting energy supply*

Factors	Influence on variations in energy supply

Physical	■ Deposits of both oil and gas, and to an extent coal, are found in a limited number of locations (geological factors influence oil and gas traps and the formation of deltaic swamps in which coal formed).
	■ Renewables are restricted by source availability:
	  – Hydropower requires a large head or high volume of water, usually linked to areas of high precipitation.
	  – Solar power ideally requires large numbers of days with strong sunlight. Tropical latitudes are especially suitable, e.g. Sahara.
	  – Wind power relies on high, constant wind speeds, which are characteristic of areas affected by westerly belts, e.g. the UK.
	  – Tidal power is restricted to a few estuaries with a very large tidal range (e.g. Bay of Fundy, River Severn).
	  – Large nuclear power stations are best located coastally on geologically and seismically stable foundations.

Economic	■ Issues of accessibility and cost of extraction influence the speed of development, for example:
	  – Onshore land-based and shallow water offshore supplies of oil and gas are developed before deep sea oil.
	  – Currently open-cast coal mining is much cheaper than deep mining.
	  – Potential HEP sites close to transmission corridors or large centres of population are likely to have been developed at the expense of remaining remote, inaccessible locations.
	■ In developing countries FDI is essential to the development of energy resources. TNCs prefer to invest in politically stable environments.
	■ Exploitation depends on current energy prices — for example, for a barrel of oil. Unconventional and deep-water oil supplies will only become viable if prices remain high (see page 59).
	■ Availability of improved and new technology can make speculative reserves economically viable.

Political	■ Countries wishing to develop nuclear electricity require permission from the International Atomic Energy Authority as the technology can be used for military purposes (hence issues with Iran in 2008–11).
	■ Many HEP sites on international rivers, such as the Omo Dam in Ethiopia, may require the agreement of all countries that share the river (a source of potential conflict on the Nile or Mekong).
	■ Government policy plays a major role in responding to the Kyoto Protocol and its successors, which place limits on greenhouse gas emissions for all developed countries:
	  – This may control the amount and type of fossil fuel used, for example low-sulphur coal or emphasis on cleaner gas.
	  – It may lead to a drive towards renewables such as wind, with a clear target for their contribution, as in the EU.
	  – It may prohibit the use of the nuclear power option; some countries, for example Sweden, currently view it as unsafe.
	  – It may determine strategies for energy security, such as avoiding unstable suppliers and volatile pathways (see *Case study 2*).
	■ The cost of R&D is a very significant factor for many renewables and therefore developing countries struggle to gain access to the technology.
	■ Public perception of what is the correct path to achieving energy security is increasingly significant. Energy conservation is an attractive option, but many of the measures require political support.

As secure energy supply is so vital to the running of nations' economies and in helping them plan for the future, there are inevitable tensions in the relationship between energy producers and consumers. The availability of energy supplies has become a **geopolitical** issue. This especially relates to fossil fuels, which currently supply over 80% of global annual energy demands.

When considering the availability of fossil fuels there are two terms of key importance.

Proved reserves are those quantities that geological and engineering information indicates with reasonable certainty can be recovered in the future from known deposits under existing economic and operating conditions.

Reserves to production (R/P) ratio refers to those reserves remaining at the end of any year divided by the production in that year. The result indicates the length of time that these remaining reserves would last if production were to continue at that level.

Coal

Coal is the most abundant and widely distributed fossil fuel resource in the world. Coal production has increased by over 1,200 million tonnes of oil equivalent since 1999. It is dominated by the Asia-Pacific region — China alone produced around 43% of the world's coal in 2009 as it strived to power its own economic expansion. The next largest producers were the USA, Australia, India, Indonesia and Russia.

In contrast to this Asian expansion, all other regions showed either static or declining production. In the case of the UK and Germany, once major coal producers, today there are only a handful of deep mines operating. Because of costs, production is concentrated in large, open-cast pits.

Table 1.3
Coal production, 2009 (million tonnes)

Country	Production (rank)		Reserves (rank)		R/P ratio in years
China	1,552.9	*(1)*	114,300	*(3)*	38
USA	539.9	*(2)*	238,308	*(1)*	245
Australia	228.0	*(3)*	76,200	*(4)*	186
India	211.5	*(4)*	58,600	*(5)*	105
Indonesia	155.3	*(5)*	4,328		17
South Africa	140.7	*(6)*	30,408	*(8)*	122
Russia	140.7	*(7)*	157,010	*(2)*	550
Poland	56.8	*(8)*	422*		14
Kazakhstan	51.8	*(9)*	31,300	*(7)***	308
Columbia	46.9	*(10)*	6,814		95

*Note how some leading producers such as Poland have low reserves.
**Ukraine has significant reserves and is ranked in 6th place above Kazakhstan.
Source: *BP Statistical Review of Energy 2010*

In terms of reserves, as Table 1.3 shows, the USA is the world leader, followed by Russia, China, Australia, India and Ukraine. A large proportion of world reserves is shared among comparatively few large countries. While the overall R/P ratio is 119 years, many of the larger producers have over 200 years — so there is no apparent threat of peak coal (see page 42 for concerns over the true amount of reserves).

Improvements in transport and technology have led to very large crude carriers (VLCCs), which can carry huge amounts of coal around the world in slurry form leading to competitive prices for imported coal such as that from Australia, South Africa or Colombia.

The controversy as to whether coal can or should make a comeback from its reputation as the dirtiest fuel is discussed in detail in Part 3.

Gas

For some, coal is too dirty, oil is past its peak, and nuclear power is too expensive and controversial. Others argue that renewables are not necessarily green, technologically feasible or economically viable. To meet the world's energy needs and cut carbon emissions, we need an abundant source of relatively clean, relatively cheap energy. Natural gas is *now* such a fuel because of the development of the technological capacity to exploit unconventional gas, most promisingly from shale deposits. Unconventional sources will augment conventionally drilled gas, which is either transmitted as gas by pipeline or transported as **LNG** by tankers (e.g. in Qatar). This will add an extra 60–70 years (until 2140) before peak gas is reached.

Horizontal well drilling, combined with hydraulic fracturing, have been developed to get at the unconventional gas more easily. In the USA, the Barnett Shale in Texas currently produces around 300 million m^3 of gas per day. Once the Marcellus Shale gas in Pennsylvania comes on stream, up to 60% of US gas will come from unconventional sources (see page 26).

Further hotspots for unconventional gas include Australia, China, South Africa and Europe (the UK, Sweden, Germany and Poland) where the big oil and gas companies are driving the exploitation. The discovery of this new source of gas has led to a fall in world gas prices. This has hit the Russian gas industry, which is based on conventional sources (see *Case study 1*).

As clean a fuel as shale gas appears to be, there are concerns that the fracturing technology may damage the environment. It is also a very water-intensive process.

Globally in 2010 there was a glut of gas — high-priced production areas such as Russia and Turkestan showed a marked decline in production as they could not compete with competitively priced LNG from Qatar or even unconventional gas.

***Table 1.4** Gas production, 2009*

Country	Production (billion m^3)	Proven reserves (trillion m^3) (rank)	
USA	593.4	6.9 (pre shale gas)	(5)
Russia	527.5	44.4	(1)
Canada	161.4	1.8	
Iran	131.2	29.6	(2)
Norway	103.5	2.0	
Qatar	89.3	25.4	(3)
China	85.2	2.5	
Saudi Arabia	77.5	7.9	(4)
Indonesia	71.9	3.2	
Uzbekistan	64.4	1.7	

Note: Turkestan also has very large reserves
Source: *BP Statistical Review of Energy 2010*

The following case study of Russia's attempt to regain superpower status based on energy — especially gas — poses interesting questions during what is currently a gas glut.

RUSSIA: GAS SUPERPOWER?

Case study 1

Russia is one of the world's major producers and **exporters** of oil and gas (world number 1 for gas exports, and world number 2 for oil exports). Together oil and gas accounted for over 20% of Russia's GDP in 2008. With the help of a tide of 'petro dollars', Russia had paid off much of its foreign debt by 2006 and had some of the world's biggest financial reserves.

Gazprom, a state-owned energy TNC, is now the world's second biggest energy company (by market value) and participates in a wide range of investment activities around the world. It controls about one third of the world's conventional gas reserves and accounts for 90% of Russia's gas production. It currently employs over 400,000 workers. In 2006 it supplied around 25% of the EU's natural gas and was the sole gas supplier to the Baltic States, Finland, Slovakia, Macedonia and Moldova.

Using the maxim 'he who pays for the pipelines calls the tune', Russia has used gas, and to a certain extent oil, as a means of energy imperialism to reassert its power and influence over former Soviet republics (FSRs). It would argue that this was purely economic, to get a fair price for its exports; others might see it as political manoeuvring.

A series of disputes

In November/December 2004, Ukraine — an FSR — had what came to be known as the Orange Revolution, in which a pro-Russian government was ousted by one in favour of links with the West. In January 2006 (the height of winter) Russia decided to quadruple gas prices for Ukraine (it argued that this was to reflect market prices). The Ukraine government refused to pay and the supply of gas was cut off, with knock-on effects on countries such as Slovakia. In March 2008 Gazprom again cut supplies to Ukraine (by 50%) — Ukraine at that time had applied to join NATO, but Gazprom claimed the dispute was over an unpaid US$1.5 billion debt for gas supplied.

The worry for European countries was that 80% of their gas supply pipelines were routed through Ukraine, with the prospect of knock-on effects on winter supplies for most of continental Europe. Russia claimed the dispute was not with the EU but it certainly had an impact in eastern Europe.

In early 2007 Azerbaijan was in dispute with Gazprom ostensibly again over pricing, so oil exports were suspended.

In 2007 Belarus cut off a transit pipe carrying Russian oil, which supplied 25% of Germany's needs and over 96% of Poland's needs, as well as substantial quantities to Hungary, Ukraine, Czech Republic and Slovakia. The dispute was again about payment for oil and gas, and rising prices, which Belarus aimed to counter by raising an import levy on all oil and gas piped through the country.

In 2009 Belarus halted the flow of Russian gas transmitted across its territory (some 20% of EU supplies) and Gazprom cut deliveries of gas by 30% into Belarus, following a further dispute about payments. Bulgaria, Moldova and Slovakia were most affected. The row was again nominally based on economics, but highlighted deteriorating relations between Belarus and Russia. (Belarus offered a home to the ousted president of Kyrgyzstan who was replaced by a pro-Russian leader and also stalled in joining a new customs union with Russia and Kazakhstan.)

There have also been pipeline interruptions to Georgian supplies, again as a result of a political dispute and brief war over the capture of South Ossetia, a Russian speaking enclave that was formerly part of Georgia.

EU concerns

These incidents have fuelled EU fears about the reliability of Russian gas supplies, on which some of its members are over dependent. There is clearly a move to diversify supplies, so that there is less reliance on Russian supplies, yet at the same time Gazprom has invited leading European energy companies to be partners. The EU has already built more terminals for the import of LNG (for example in Milford Haven in south Wales)

Sergey Kamshylin/Fotolia

Figure 1.5
A gas pumping
station in Ukraine

and is beginning to exploit its own unconventional supplies. It also uses the ultra reliable Norwegian supplies to a greater extent, meaning Gazprom had lost over a third of its European market by 2009.

A further concern voiced by the IEA (International Energy Agency) is that Gazprom may not have enough gas to supply Europe over the next decade, as the main fields in western Siberia, which supply 75% of Russian gas, are declining and are already augmented by Turkmenistan gas, with new production only beginning in the Arctic, and in the far east in Sakhalin. With falling prices for gas it is debatable whether exploitation of the Shtokman gas field in the Barents Sea will be economically viable (complex, high-cost location and an inaccessible area — it has been postponed until 2014). Some of the Sakhalin gas was to be exported as LNG to the west coast of the USA, but with the availability of shale gas in the USA there is no longer a need for this.

It would seem, therefore, that the glut of gas and the falling prices put Russia's aim to be an energy superpower on hold, but only temporarily. With the shift of production further east there is hope that Russia could sell gas to China, which aims for natural gas to account for at least 10% of its energy mix by 2015, and other Asian countries. However, there is much competition from Australian LNG, and also supplies from central Asia, while indigenous shale gas will become more competitive than pricey Russian supplies. In 2006 Russia began constructing the ESPO pipeline from east Siberia to Vladivostok and to northern China, to provide a new energy pathway and give Japan and China a new source of supplies.

The EU is seeking to avoid any Russian unreliability of supply by building new pipelines to bypass the Russian East-West ones.

New pipelines

In 2009 the Nabucco pipeline was inaugurated. It will connect Europe to gas-rich central Asia via the Balkans, Turkey and the Caucasus, with support from all the transit countries, at a cost of €200 million. The aim is to carry 30 billion m^3 of gas a year (about 20% of what Russia currently exports to Europe), beginning in 2015. Supplies could come from Azerbaijan and Iraq's Kurdish region, with the possibility of surplus gas from Turkmenistan (currently all this gas is sold to Russia) subject to the building of a trans-Caspian sea extension.

A further EU scheme is the ITGI , which brings Azeri gas to Italy via Turkey and Greece, being linked with a possible new trans-Adriatic pipeline.

Russia has always claimed that its disputes about gas are not about relationships with the EU. It therefore plans to go ahead with Nord Stream (costing €13 billion) and South Stream (costing €20 billion) which will solve any Ukraine/Belarus transit issues, thus ensuring a more reliable supply. It is also pressing on with the Opal pipeline to connect Nord Stream to an existing transit point on the Czech–German border. The projects make sense, but with the world gas glut will there be enough demand to make them profitable?

The real issue keeping Gazprom from domination is the high costs of Russian gas exploitation, with falling production on the accessible fields, as well as the poor financial state of Gazprom, with debts in 2009 of over $40 billion. Many Western oil companies cooperate with Gazprom but find the Russian economic climate difficult.

In the future, EU legislation may outlaw any monopoly agreements that currently exist (see *Case study 4*). Currently Russia needs European markets for its gas, probably more than the EU needs highly priced Russian gas, but the situation could well change in the future.

With the success of OPEC in maintaining the price of oil there has been a move, largely driven by Russia, to set up a similar cartel for gas. In 2008 the initial negotiations for the formation of GECF (Gas Exporting Countries Forum) took place. Fifteen countries were involved including Russia, Iran, Algeria, Qatar, Indonesia, Malaysia, Egypt, UAE, Trinidad & Tobago and Nigeria. The problem was that it was formed against a backdrop of falling consumption and prices no longer linked to oil prices, in response to the development of unconventional supplies. Also, as gas is largely used for electricity generation and heating, it is just one of a number of alternatives, so if prices rose too high, alternatives would be sought. Moreover, many of the producing countries rely on energy TNCs to provide the infrastructure for the complex process of gas liquefaction, so have limited control over planning. At present the potential members have very divergent interests and very different approaches to gas production, so there is no chance of GECF developing the power of OPEC.

This case study shows that in terms of gas the Russian geopolitical dream is just about over. There is, however, still oil (Russia supplies over 50% of EU markets in Hungary, Slovakia, Finland, Poland and Czech Republic and over 25% in the Netherlands, Belgium and Germany).

Question

Evaluate the evidence that Russia is using its huge resources of energy to enhance its superpower status.

Guidance

Look at arguments for and against this statement.

For:
- Activities of Gazprom (see **www.gazprom.com**)
- Disputes with FSRs — were they economic or political?

Against:
- New EU pipelines for improving access
- High cost of Russian gas

Conclusion:
- Maybe before the gas glut there was some evidence of this.

Oil

Oil is fundamental to twenty-first century living; oil-producing countries with substantial reserves can wield considerable political and economic power. The major oil producing countries founded the cartel OPEC in 1960 after US legislation imposed quotas on Venezuelan and Persian Gulf oil imports in favour of supporting Canadian and Mexican oil industries. Currently there are 12 members (Indonesia left in 2009 as it became an oil importing country).

OPEC's stated aim is:

> …to coordinate and unify petroleum policies of member countries and ensure the establishment of oil markets in order to secure an efficient, economic and regular supply of petroleum to users, but at the same time a steady income to producers and fair return on capital to those investing in the oil industry.

Inevitably with such a powerful geopolitical weapon, some decisions on adjusting production to maintain prices can seem political. High oil prices have a major impact on global economic wellbeing. For example, in response to the world recession in 2009 OPEC members cut production by over 7% in order to maintain the oil price — it had declined to under $40 from a high of $147. As a result, the price climbed to a fluctuating $60–80 a barrel.

Figure 1.6 shows the world's proven oil reserves and identifies the 12 OPEC members who are concentrated in the Middle East and produce nearly half of the world's crude oil. Together they hold 77.2% of total reserves with a very favourable R/P ratio of 85 years, which contrasts sharply with non-OPEC producers in which

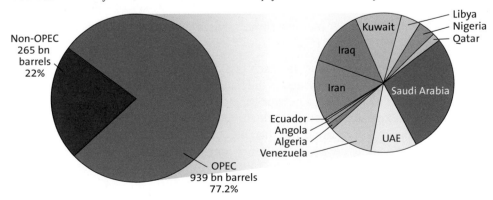

Figure 1.6
Proven oil reserves

the R/P ratio is 15 years. The country with the most significant reserves outside the Middle East is Venezuela, with 13%. The president, Hugo Chavez, known for his anti-US views, wishes OPEC to take a stronger political and geopolitical role by keeping the price of oil high at $80–100, with affluent countries such as the USA paying the most, to combat climate change, and the world's poorest countries being protected from the high prices.

Table 1.5 shows the main oil producers and their proven reserves.

Table 1.5
Oil production, 2009

Country	Production (thousands of barrels daily)	Proven oil reserves (thousand million barrels) (rank)		R/P ratio in years
Russian Federation	10,032	74.2	(7)	20.3
Saudi Arabia*	9,713	264.6	(1)	74.6
USA	7,196	28.4	(12)	10.8
Iran*	4,216	137.6	(3)	98.4
China	3,790	14.8	(13)	10.7
Canada	3,212	33.2	(11)	28.7 ‡
Mexico*	2,979	11.7	(16)	10.8
UAE*	2,599	97.8	(6)	100+
Kuwait*	2,482	101.5	(5)	100+
Iraq*	2,481	115.0	(4)	100+
Venezuela*	2,437	172.3	(2)	100+
Norway	2,342	7.1	(17)	8.3
Nigeria*	2,061	37.2	(10)	49.5
Brazil	2,029	12.9	(14)	17.4
Algeria*	1,811	12.2	(15)	18.5
Kazakhstan	1,682	39.8	(9)	64.7
Libya*	1,652	44.3	(8)	73.4

*OPEC member
‡ (excludes tar sands)
Source: *BP Statistical Review 2010*

Table 1.5 suggests that the Middle East is likely to dominate oil production for the foreseeable future. However, with the exploitation of more unconventional sources of oil (extreme oil, tar sands, shales etc.) and the discovery of new oilfields in Africa and South America, there may be some adjustments to these data in the next 10 years.

Technological requirements for deep-sea offshore drilling are cutting edge, so the exploitation process is concentrated in the hands of energy TNCs. As key consumers are rarely the main producers, a network of pipelines and oil tanker routes is vital for trade to occur. Exporting regions include Russia, the Middle East and parts of Africa, while the main importing areas are the OECD countries in Europe, Asia and Asia-Pacific. Disruption to either tanker routes or pipelines can cause major problems in oil supplies — an ever-present concern for large consumers of imported oil.

Energy security

Energy security can be defined as having uninterrupted access to reliable sources of supply, at an affordable price, with extraction and utilisation not having an undue impact on the environment. Inevitably there are spatial variations (see Table 1.2), but there are also variations over time. With the advent of new discoveries, or new technologies, which either make new sources feasible (for example wave power) or unconventional resources economically viable (for example gas from shale), energy security has improved in some countries (see *Case study 2*). In other cases, security can get worse with rising consumption (e.g. China and India) or over-reliance on importation, whereby disruption of supplies can take place either at source or along pathways — for example as a result of piracy on tanker routes or pipeline disputes (see *Case study 1*).

Risks to energy security can be classified as follows:
- **Physical** — Reserves are exhausted (for example North Sea oil and gas may be past its peak) or supply lines and exploitation may be disrupted by natural hazards such as earthquakes or hurricanes (Gulf of Mexico oil supplies).
- **Environmental** — Green protesters are increasingly vociferous about environmental damage. Global examples include open-cast coal mining, or exploitation in sensitive areas such as oil in the Arctic, as well as rising levels of greenhouse gases from the use of fossil fuels, and the building of nuclear power stations. Locally nimbyism can put a brake on the development of potentially eco-friendly developments such as wind farms, causing a 'short-term energy supply gap' for many countries, in the drive to replace dirty fuels.
- **Economic** — The fluctuations in the price of oil per barrel can have an adverse impact on both energy development (see page 59) and energy costs. Sudden rises in the cost of energy (for example oil in 2008) or exhaustion of domestic supplies requiring more imports of higher priced energy both cause increased risk to security.
- **Geopolitical** — These risks include political instability in energy producing regions, e.g. Middle East wars, disputes or conflicts over ownership (rights of native peoples such as the Nenets in Russian Arctic) and disputes over transmission (the issue of Russia's pipelines).

Figure 1.7 maps the degree of global energy security risk in 2006.

The Energy Security Index

Energy security is a complex concept, so it is difficult to measure. The Energy Security Index (ESI) uses three components to classify the countries in Figure 1.7.

Availability — the production and proven reserves of each country's domestic oil and gas supplies and its level of reliance on imports.

Diversity — the range of energy sources in the energy mix each country uses to meet its demand.

Intensity — the degree to which the economy of each country is dependent on oil and gas (the resources where peak in production is most likely to be imminent).

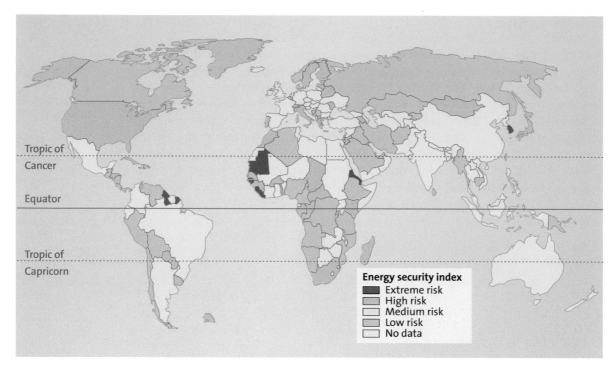

Figure 1.7
World map showing energy security risk

ESI values range from 0–10. Figure 1.7 shows four groups.

Extreme risk — ESI values less than 2.5. With the exception of the Korean Republic (which is taking huge steps to secure its oil supplies), the countries are developing countries in west Africa (e.g. The Gambia) or in Latin America (e.g. the Guyanas). So important are oil and gas for Korea that the government has allocated money to the Korean National Oil Corporation to buy up exploration and production companies. In August 2010, for example, a hostile bid was launched to buy up Dana Petroleum, a North Sea exploration company.

High risk — ESI values 2.5–5.0. As the map shows, these countries are scattered across the globe and include developed countries such as the USA (now improving), Japan and (since 2010) China, whose demand has risen dramatically.

Medium risk — ESI values 5.0–7.5. A widespread category, including most European countries, many parts of Africa where demand is low and HEP supplies are enormous, as well as Brazil, Argentina and Australasia. Brazil is an unusual case — as well as making recent discoveries of offshore deep-water oil supplies, it has invested heavily in ethanol development to improve its security.

Low risk — ESI values greater than 7.5. This category includes many oil and gas producing countries such as Canada, Norway, Venezuela and Peru, all of which have surplus to export, as well as low-consuming African countries such as Nigeria, Ghana and Angola, where oil has only recently been discovered and so is not fully exploited. It also includes the Middle Eastern countries that form the core of OPEC and Russia.

USA: A COUNTRY WITH A HIGH ENERGY SECURITY RISK?

The USA consumes huge quantities of energy, largely for transport, heating and electricity generation. As oil and natural gas are the major sources of fuel for these needs, it is inevitable that in spite of having considerable indigenous resources, the USA relies on imported fuel. This is a major feature of a country with high levels of energy consumption per capita (at 12,187 kWh per year, about six times the world average) although in terms of total consumption it is being hotly pursued by China.

In 2009 the USA consumed over 20% of the world's global energy supplies and around 23% of the world's oil. In 2006 60% of all its primary energy requirements, including a staggering 80% of its oil, was imported.

A drive to import less arose from a number of factors. The oil 'shocks' of 1973 and 1979 galvanised US politicians in trying to reduce the country's dependence on imported oil, but since prices crashed in 1986 there has been an almost continuous increase in amounts of imported oil.

However, over the last decade the US government has refocused on finding alternatives to imported oil and gas. The latest 2008 oil shock, fuelled by huge demand in emerging markets such as China when prices rose to nearly $150 per barrel, was far-reaching. It led to a huge deficit in the American economy — net imports of oil and gas cost $416 billion, around 60% of the deficit.

As peak oil approaches, prices are predicted by some analysts to rise to $200, perhaps as early as 2015. This threat to the gas guzzling US way of life is combined with the Middle Eastern terror threat. In the words of President Bush in 2006, 'America must end its dependence on oil; when you're hooked on oil from the Middle East, it means you've got an economic security issue combined with a national security issue'. Moreover, several states, including California, began to experience electricity shortages as a result of poor planning in the fragmented privatised sector.

Indigenous oil

Efforts have been increased to exploit US and other 'friendly' neighbouring oil supplies in Canada and Mexico. With prices high, 'extreme oil' reworked from ageing fields in the Great Plains (using new technologies such as gas and steam injection), Athabasca tar sands, and Green River oil shales in Wyoming have become economic. The high prices have also encouraged deep-water offshore exploration, such as on the Tiber field (4–6 million barrels) in the Gulf of Mexico (temporarily halted by the enormous oil spill from the *Deep Water Horizon* rig in Spring 2010). President Bush gave permission for further exploitation of Arctic resources in the environmentally sensitive area of the Alaskan National Wildlife Refuge (ANWR), although President Obama has restricted this to offshore areas.

Figure 1.8 shows what a difference all these measures could make to American oil imports.

Figure 1.8
US oil imports

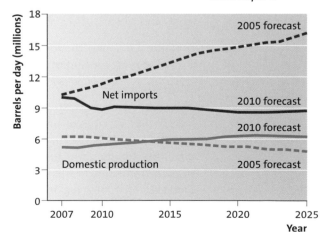

The dash for shale gas

The other main trend has been in the exploitation of shale gas in the Marcellus Shale of Pennsylvania and the Barnett Shale of Texas. Two well-tried technologies — drilling horizontally instead of vertically and releasing the gas by hydraulic fracturing (injecting water, sand and chemicals at high pressure into the rock) make this possible.

There are environmental concerns because of the excessive use of water, the potential for water pollution and also the uncontrolled spread of numerous little wells. However, the pressure to exploit indigenous unconventional gas is huge and in general it is a much cleaner fuel than coal and oil in terms of greenhouse gases. By 2010 the USA had become the largest gas producer in the world, at nearly 700 billion m³ and the price for this gas had fallen from $13 million BTU to less than $5 million BTU. Estimates before shale gas suggested that nearly 30% of gas (LNG) would be imported; now the figure is only 9%.

Figure 1.9
Location of US shale gas resources

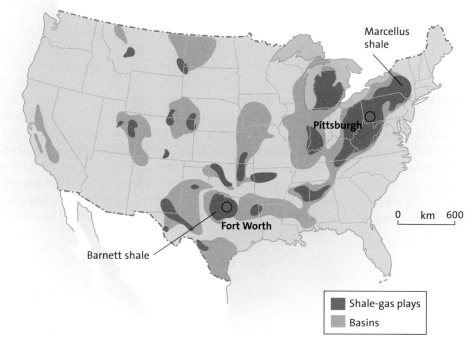

Figure 1.9 shows the location of the unconventional gas reserves, with the 'plays' being the most concentrated areas.

Alternatives

The USA has also invested heavily in biofuels, using maize, which now contributes significant quantities of subsidised ethanol and biodiesel to the US economy. Industrial biotech has the potential to replace oil with agricultural feedstocks to produce chemicals, plastics and fuels. The Federal Renewable Food Standard passed in 2005 requires refineries to blend 36 billion gallons of renewable fuel such as ethanol or biodiesel into motor fuel by 2012 (i.e. 17% of consumption).

In response to climate change, President Obama has promoted numerous new measures to start investment in renewable energy such as wind or solar, for which states such as California have great potential.

The nuclear option

Currently the USA has 65 nuclear power stations, largely concentrated in the densely populated northeast. There have been no new nuclear power stations built since the Three Mile Island accident in 1979, which vaporised public support for nuclear power. President Obama proposed a new generation of safe, clear nuclear power plants in his State of the Nation Address, to reduce carbon emission and to promote US 'energy independence'. He announced $8.3 billion in loan guarantees to build two new nuclear power stations in Georgia and a further 16 proposals are in the pipeline.

Currently nuclear power provides 20% of US electricity, with coal at nearly 50%. Remember that the USA has the most reserves of coal in the world. Public support for nuclear power can be described as widespread but shallow, with some states passing bans on nuclear plant building, and others seriously considering them. A nuclear power renaissance is enhanced by its cleanliness, i.e. its low contribution to greenhouse gases, but other issues are very significant (see Part 5) such as vulnerability to terrorism, radioactive waste, and the huge amount of funding and expertise required by privatised electricity companies to build nuclear power stations. With the dash to unconventional gas, the resurgence of nuclear power hinges on public support.

Falling demand

Clearly the impact of the 2009 recession led to a huge decrease in consumption both by domestic and industrial users. However, when the USA recovers, will this drop in demand be sustained? There is no doubt that changes have taken place in the car and aircraft industries with lighter vehicles and planes (e.g. the 787 Dreamliner) with more efficient engines. Hybrid and electric cars are also gaining in significance. America's consumption of petrol and the number of miles driven peaked in 2007, with improved developments in public transport in major cities and the revival of city centre as opposed to mall shopping. However, in 2010 the price of oil per barrel dropped back to around $80, with petrol down to $3 per American gallon, and recovery and declining unemployment occurring. Will there be a lasting change in consumer behaviour and maintenance of a decreased demand for fuel?

2 Using case studies

Questions

1 Evaluate the strengths and weaknesses of options available to the USA to improve its energy security. Which do you consider is likely to be the most significant driver and why?
2 Look at another country with a high energy security risk, such as China or Japan, and list the strategies that it is using to improve the situation.

Guidance

1 In your evaluation look at environmental, economic and political pros and cons.
2 Look at securing fossil fuel supplies, the nuclear option, use of HEP and other renewables, as well as conservation and increased energy efficiency.

Patterns of consumption

Global energy use

All scenarios predict a steadily rising consumption of energy in spite of a downturn brought about by the global recession. In 2008–09 there was a fall of 1.1% in overall primary energy consumption, which dropped back to 2007 levels. The trends also show that the rise in global demand will be driven by non-OECD countries (especially the emergent superpower of China and others such as Brazil, India, Mexico and the Middle Eastern states). A combination of increasing population growth, rising living standards and rapid industrial expansion has led to an enormous demand for energy in these areas, which in some cases could rise by over 90% between 2007 and 2030 (according to the International Energy Agency's 'business as usual' scenario — Figure 2.1).

Even in the IEA '450 green growth' scenario, demand only begins to fall below 10,000 Mtoe by around 2025. Note that consumption post 2010 in OECD countries (the richest nations) remains static.

As a nation develops economically, both its primary energy consumption and its use of petroleum and electricity (secondary sources) rise.

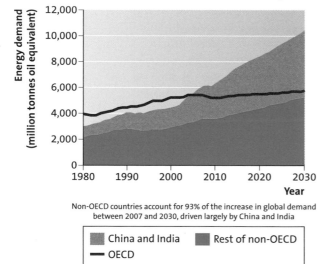

Non-OECD countries account for 93% of the increase in global demand between 2007 and 2030, driven largely by China and India

Legend:
■ China and India   ■ Rest of non-OECD
— OECD

Figure 2.1
World primary energy demand in the IEA reference scenario ('business as usual')

Figure 2.2 is a scatter graph that attempts to correlate energy consumption per capita with GDP per capita (Table 2.1). You can also visit **www.gapminder.org** to produce graphs linking energy use to the human development index (HDI).

Put a best-fit line through the scatter graph and you should find a positive correlation. The higher the personal spending power and export-led wealth generated, the greater the demands for energy. There are some anomalies — for example, oil-rich states of the Middle East have access to huge amounts of cheap oil-based energy, but some (e.g. Saudi Arabia) have very unequal societies, so the overall **per capita** income is only mid-range.

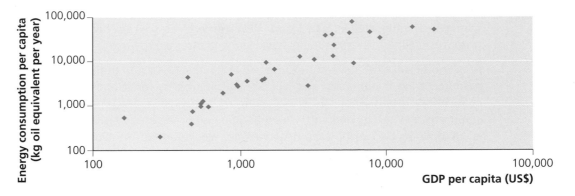

Figure 2.2 *Energy consumption per capita*

Table 2.1

Country	Total energy consumption per capita per annum, 2003 (kgoe per year)	GDP per capita (US$)	Country	Total energy consumption per capita per annum, 2003 (kgoe per year)	GDP per capita (US$)
Australia	5,723.3	42,279	Kuwait	9,076	33,861
Bangladesh	160.9	551	Libya	3,203.2	10,100
Bolivia	503.8	1,292	Mexico	1,533.2	9,898
Chile	1,652.2	9,013	Norway	5,933.6	79,090
China	1,138.3	3,687	Peru	431.5	4,345
Costa Rica	879.9	5,046	Qatar	21,395.8	52,240
Cuba	935.1	2,864	Romania	1,784	6,932
Czech Republic	4,319.3	13,877	Tanzania	464.9	394
Ethiopia	277.9	203	Thailand	1,405.7	3,810
Germany	4,203.1	41,230	Tunisia	933.3	2,990
Hungary	2,595.2	12,868	Ukraine	2,968	2,879
India	512.4	1,122	United Arab Emirates	10,538.7	58,827
Indonesia	757.4	1,934	UK	3,918.1	38,850
Jamaica	1,545.1	4,014	USA	7,794.8	46,155
Kenya	481.2	759	Vietnam	539.4	998
South Korea	4,346.5	23,340	Zambia	600.6	940

A model of energy transition

The modernisation theory (developed by Rostow) identifies a number of stages that countries pass through as they develop economically from traditional societies, through economic take-off, to high mass consumption.

An energy transition model can be developed (see Figure 2.3) to show how energy demand increases over time, and how the actual sources of energy (the energy mix) can vary as a country develops. For example, in traditional societies such as the rural areas of Niger, Burkina Faso or Nepal there are largely subsistence economies, with low energy consumption, based on burning fuelwood and other biomass. Trading and infrastructural development, as well as diversification of primary activity (forestry, mining), will lead to greater energy use.

At economic take-off stage, for example China in the 1990s and India in the 2000s, manufacturing is dominant and will often be driven by fossil fuels — especially coal. Rapid urbanisation, as is occurring in many NICs, leads to rising living standards and energy demand, with a broadening of the energy mix. While China is still building two coal-powered stations per week and has an overall coal-based economy, with endemic pollution problems, it has policies to widen its fuel mix to include more gas, HEP, nuclear and renewables.

Advanced economies such as the USA and Japan rely on tertiary/quaternary industries that depend heavily on secondary energy supplies, such as electricity, which can be obtained from a wide mix of fuels. Adherence to the Kyoto Protocol (and its successors) and concerns over peak oil have meant that many advanced countries have been forced into more sustainable strategies, with decreased use of fossil fuels (especially coal and oil).

Some developing countries with access to lower-cost Chinese and Indian renewable technologies (solar, wind and hydro) are able to move towards sustainable green development and largely bypass coal-based industrialisation. This applies to countries such as Kenya and Ethiopia.

Figure 2.3
An energy transition model

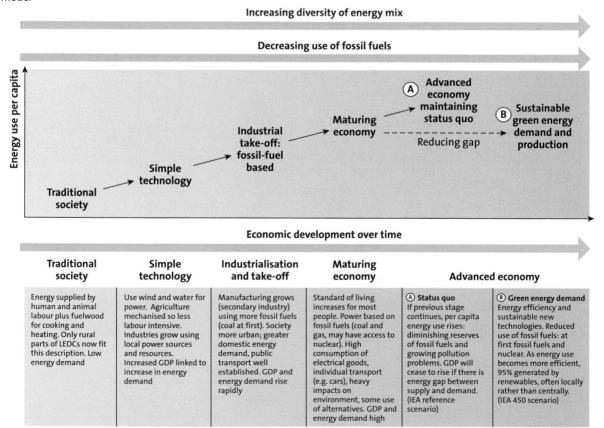

Traditional society	Simple technology	Industrialisation and take-off	Maturing economy	Advanced economy	
Energy supplied by human and animal labour plus fuelwood for cooking and heating. Only rural parts of LEDCs now fit this description. Low energy demand	Use wind and water for power. Agriculture mechanised so less labour intensive. Industries grow using local power sources and resources. Increased GDP linked to increase in energy demand	Manufacturing grows (secondary industry) using more fossil fuels (coal at first). Society more urban; greater domestic energy demand, public transport well established. GDP and energy demand rise rapidly	Standard of living increases for most people. Power based on fossil fuels (coal and gas, may have access to nuclear). High consumption of electrical goods, individual transport (e.g. cars), heavy impacts on environment, some use of alternatives. GDP and energy demand high	ⒶStatus quo If previous stage continues, per capita energy use rises: diminishing reserves of fossil fuels and growing pollution problems. GDP will cease to rise if there is energy gap between supply and demand. (IEA reference scenario)	ⒷGreen energy demand Energy efficiency and sustainable new technologies. Reduced use of fossil fuels: at first fossil fuels and nuclear. As energy use becomes more efficient, 95% generated by renewables, often locally rather than centrally. (IEA 450 scenario)

Regional consumption patterns

A further feature of consumption is the variation in regional consumption patterns, as shown in Figure 2.4.

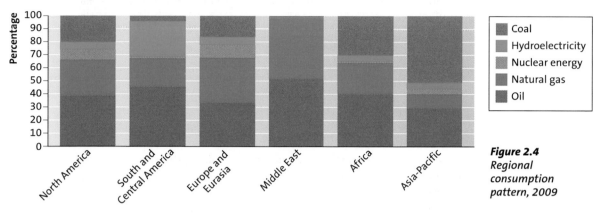

Figure 2.4
Regional consumption pattern, 2009

While oil remains the world's dominant fuel, it has lost market share over the last decade — both globally and regionally. Natural gas now has the leading market share in Europe and Eurasia, and it is a close second to oil in the Middle East. Only in the Asia-Pacific region is coal the dominant source of energy. The relative importance of hydroelectricity is greatest in South and Central America, although it will grow in Africa, where many large schemes are underway in the Congo Basin, and the Ethiopian Highlands. Nuclear energy is largely restricted to Eurasia and North America because there is a cross-over of technology with the development of nuclear weapons and existing nuclear powers are keen to restrict its spread.

What is very clear is that if a 'business as usual' scenario prevails, the world will continue to rely on fossil fuels. As Figure 2.5 shows, fossil fuels are likely to account for 77% of the increase in world primary energy demand in the next 20 years. Note the marked contrasts between OECD and non-OECD countries.

In the Examination advice section on analysing graphs (page 110) you have the opportunity to compare future trends in primary and secondary energy consumption. The shift towards more sustainable energy futures is happening at a very slow pace, especially in primary energy consumption (page 93).

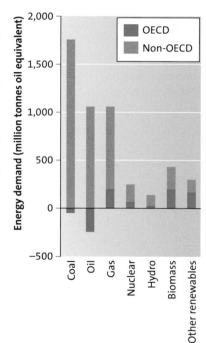

Figure 2.5
Change in primary energy demand for the IEA reference ('business as usual') scenario, 2007–30

Energy mix

The energy mix of a country is the particular combination of energy sources used within it for production of power and consumption. The mix can vary hugely depending on a number of factors:

■ **Level of availability** and the type and range of indigenous resources are major factors. The oil crises and subsequent rises in price in 1973 and 1979 led many countries to look at developing domestic energy sources. This is seen in the USA

(*Case study 2*), which has considerably enhanced its energy security by developing deep-sea offshore oil, and Alaskan oil in the Arctic National Wildlife Reserve. Clearly, environmental, sociocultural and economic impacts need weighing up against the political advantage. Some countries are physically more suited for developing one resource, for example Ethiopia for HEP or Iceland for geothermal power, while others have a very diverse base (e.g. the UK or Denmark).

■ **Security of supplies** beyond indigenous resources is another consideration. *Case study 4* of the changing energy mix in the EU emphasises the need to link up with friendly countries that can guarantee security of supply. Thus the EU has moved to cut dependence on Russian oil and gas supplies and look instead to countries such as Norway. There are also global issues of security of supplies of oil that may lead to a shift in sources for electricity generation, or even transport.

■ **Level of development** impacts on the energy mix of a country because developing nations may lack the technological expertise to develop resources or the infrastructure to transport them. For example, building LNG terminals or nuclear power stations requires foreign direct investment (FDI) from transnational specialist energy corporations. The populations in developing countries do not have the money to buy alternative energy sources, so rely heavily on biomass such as fuelwood as their main energy supply. Therefore, **affordability** of supplies is a key issue.

■ **National and international legislation** can also have a huge impact on the energy mix. For example as a result of the Kyoto Protocol, and subsequent emissions targets, many countries have shifted towards cleaner fossil fuels such as gas, and renewables. Government taxation policies can favour the use of one form of energy at the expense of others — for example, by reducing tax on cars with lower emissions, placing a carbon tax on the use of fossil fuels or promoting renewable energy through community grants and subsidies (e.g. Germany). As *Case study 4* shows, the EU is even using litigation to prevent the development of supply monopolies and thus ensure competitiveness.

■ **Cultural preferences** can also slow the adoption of modern, cleaner energy resources. In countries such as the UK and Poland many people prefer the concept of a 'living fire' — an open coal or wood fire rather than a gas or electric one.

Case study 3 ENERGY MIX IN THE ASIA-PACIFIC PARTNERSHIP

This case study looks at influences on the energy mix of the six nations of the Asia-Pacific Partnership and the need for clean energy development. These nations already account for about half the world's energy consumption and carbon dioxide emissions, and without action are forecast to double their totals on both counts by as early as 2025. They consume 44.4% of the world's energy, are home to 44.9% of the population and produce 51.6% of the world's carbon dioxide emissions. These nations plan to use a range of technologies to cut carbon emissions, for example:
■ carbon capture and sequestration of coal
■ clean coal technology (pulverisation and gasification)
■ enhanced energy efficiency in buildings and cars

They also plan to generate more electricity from wind, solar and nuclear power and to exploit unconventional shale gas supplies.

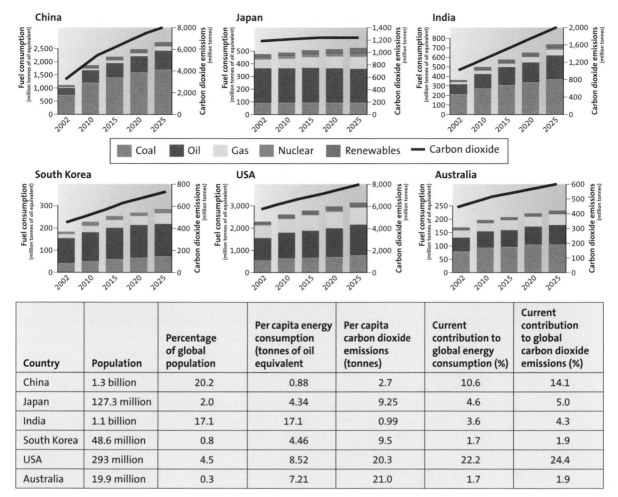

Figure 2.6 *Comparing the energy mix of the six countries in the Asia-Pacific partnership*

Country	Population	Percentage of global population	Per capita energy consumption (tonnes of oil equivalent	Per capita carbon dioxide emissions (tonnes)	Current contribution to global energy consumption (%)	Current contribution to global carbon dioxide emissions (%)
China	1.3 billion	20.2	0.88	2.7	10.6	14.1
Japan	127.3 million	2.0	4.34	9.25	4.6	5.0
India	1.1 billion	17.1	17.1	0.99	3.6	4.3
South Korea	48.6 million	0.8	4.46	9.5	1.7	1.9
USA	293 million	4.5	8.52	20.3	22.2	24.4
Australia	19.9 million	0.3	7.21	21.0	1.7	1.9

The partnership has signed agreements for technology transfer and sharing new technologies. With these six countries at the forefront of new energy technology this planned synergy could prove very successful.

Figure 2.6 summarises the profiles of the six countries in terms of their planned consumption and energy mix. Note in all cases the overwhelming reliance on fossil fuels both currently and in the future, with only Japan beginning to show a diminishing reliance (brought about by a slight planned decline in coal use). Overall there is increasing use of cleaner fuels if gas is included, especially in the USA and South Korea.

Country profiles
China
China has a huge projected growth in energy demand as it develops economically (it currently accounts for 33% of the growth in global demand for oil, 60% of which has to be imported, as China's own oilfields have peaked). China is by far the world's biggest producer and consumer of coal — 70% of electricity is produced using coal, with dire consequences for atmospheric pollution, hence China's move towards renewables such as hydroelectricity, wind and solar power (see *Case study 21*). It is also expanding its

nuclear energy programme and exploration for gas resources, both conventional and unconventional.

China has a problem of regional imbalance, with many of the supplies of oil and gas found in the north and west and most power use concentrated in the coastal core in the east and south, hence the desire to import oil and LNG from a range of sources. China's energy insecurity rating is medium, moving to high (energy dependency 15%). It is strengthening its army to guard supply routes, and is actively involved in developing oil supplies in Africa and South America.

Japan

Japan was until recently the largest economy in Asia, but it has the smallest energy reserves of any major economy with 99% of its oil and gas imported. It has an energy dependency of 80%, with high insecurity. Japan has been seeking to reduce its dependence on the Middle East by engaging with Russia and FSRs in central Asia. It also has a large nuclear programme, producing 25% of its electricity, and is investing in renewables, especially HEP. Energy efficiency measures have led to a flattening of demand.

India

India has escalating demands as a result of its rapid development — between 1973 and 2005 the country's energy consumption grew by over 300%. The fuelwood/biomass sector has declined. In response to diminishing indigenous oil reserves India has embarked on an extensive programme of developing renewables, especially solar and wind, extended its HEP projects (Sardar Sarovar and Narmada) and begun an extensive nuclear programme. Note how its demand is much lower than China but that demand is rising dramatically, relying on a similarly diversified mix. India's energy insecurity is medium, but likely to become high in the absence of indigenous fossil fuel resources except coal.

South Korea

South Korea is extremely energy insecure with no real indigenous sources. It is heavily reliant on imported oil and also coal and gas to fuel its rapid industrial development. It is developing an extensive nuclear power programme and has taken steps to establish friendly links with fossil fuel providers round the world in a bid to secure supplies.

USA

The USA has moved towards improved energy security, essentially by developing new indigenous supplies of unconventional oil and gas, as well as states investing in a range of renewable technologies via Federal loans. Electricity privatisation means that energy future planning can be piecemeal, and uncoordinated between states (see *Case Study 2*).

Australia

Australia is relatively well endowed with indigenous resources including coal, limited oil, gas, HEP, uranium and solar power, which means that its energy insecurity is low–medium, especially as it has a low population of around 20 million people. Unusually it does not have a nuclear programme and was very late in signing up for Kyoto. With nearly half of its energy coming from coal, investment in clean coal technology will be vital for a 'cleaner future'. Australia is predicted to reach a surplus of gas, with new gas fields in Western Australia and also shale gas potential, and is a significant exporter of LNG to other Asia-Pacific countries.

THE NEW EU ENERGY STRATEGY

The EU is the world's second largest energy market, with over 450 million consumers. Cooperation between member states is vital to deal with the key energy issues facing the EU in the next 25 years. In 2006 a green paper entitled 'Balancing sustainable development, competitiveness and security of supply' outlined the EU strategy. The policy has three main objectives:

Sustainability

- Developing competitive renewable sources of energy — the 'Renewable energy road map' (2008) suggests a binding agreement of 20% renewables in the EU's energy mix by 2020 and a binding minimum target for biofuels for transport of 10% by 2020. The target for 2010 was a 12% share of renewable energy in its overall mix, but although renewable energy consumption had increased by 55% since 2000, its share was unlikely to exceed 10%.
- Developing other low-carbon energy sources and carriers, in order to fulfil climate change targets.
- Curbing energy demand by renewed energy efficiency efforts.
- Playing a leading role in global efforts to mitigate the impacts of climate change and improve on quality as a cooperative strategy, for example the EU Emissions Trading Scheme.

Competitiveness

- Ensuring that energy market decisions bring benefits to consumers and to the economy as a whole by developing competitive energy markets, while stimulating investment in clean energy production and energy efficiency.
- Mitigating the impact of higher international energy prices on the EU economy and its citizens.
- Keeping Europe at the cutting edge of energy technologies, for example commercialised carbon sequestration and clean coal technologies.

Security of supplies

- Tackling its rising dependence on imported energy. Unless the EU makes its domestic energy more competitive in the next 20 years, around 70% of its energy requirements will need to be imported, compared with 51% in 2010.
- An integrated approach that reduces demand and diversifies the EU's energy mix, with greater use of competitive indigenous and renewable energy.
- Diversifying sources and routes of supply of imported energy (the issue of the European over-dependence on Russian oil and gas is illustrated by Figure 2.7).
- Creating a framework that will stimulate adequate investment in new-generation nuclear and gas-fired power stations to replace ageing dirty coal-fired power stations such as Buildwas in Shropshire.

Figure 2.7
Russia's power-hungry customers: (a) percentage of EU countries' oil supplies originating in Russia, 2006; (b) Russia's European gas exports by consumer, 2006

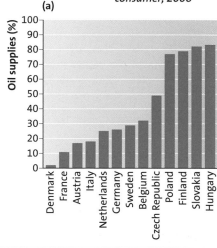

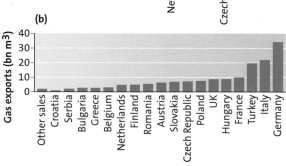

- Better equipping the EU to cope (as a united superpower) when energy supply emergencies arise.
- Improving infrastructure for EU countries seeking access to global resources (e.g. building LNG terminals).
- Making sure that all citizens and businesses, including those in the unconnected peripheral 'energy islands' of Ireland, the Baltic States and Finland have equitable access to diverse supplies of energy.

Note that although minimal levels of secure and low-carbon sources will be agreed between all member states, the freedom of member states to choose between different energy sources to achieve an appropriate mix will be guaranteed.

Reserves and imports

Europe is comparatively poorly endowed with proved reserves of energy resources and their distribution is dispersed. In 2008 the EU imported most of its oil: 16% from Norway,

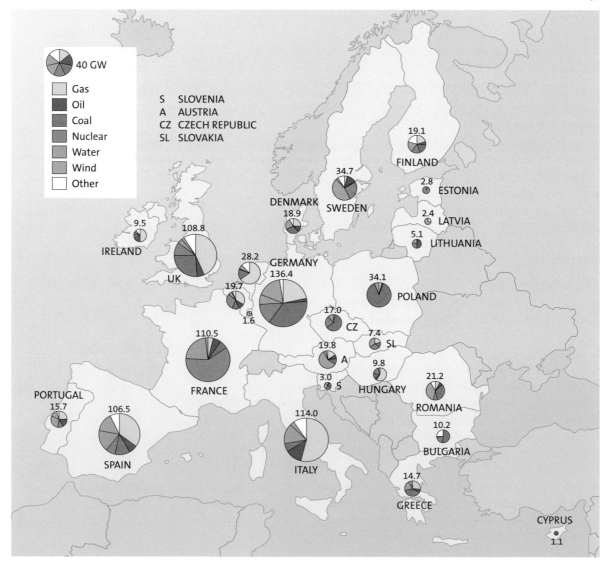

Figure 2.8
EU electricity production, 2007

38% from OPEC and around 33% from Russia. The gas sector is more optimistic — domestic production, mainly in the Netherlands, satisfied around 40% of consumer needs in 2006. Gas is imported from four big suppliers, Russia 42%, Norway 24%, Algeria 18% and Qatar (LNG) 5%.

Coal reserves are widely dispersed, with substantial hard coal supplies in Poland, and lignite in Germany, Czech Republic, Poland, Bulgaria, Greece and Hungary. Sources of coal imports include Russia 26%, South Africa 25% and Australia 13%. Imports satisfy around 13% of total demand.

Although overall energy import dependence in the EU is high and currently increasing, the situation varies significantly from county to country. Denmark is the sole EU country that is completely import-independent; others, such as the UK and Poland, have quite low import dependencies (around 20%). Some countries have import dependency ratios exceeding 80% (Italy, Ireland, Portugal and Spain), with certain island states like Malta and Cyprus nearly 100% dependent (except for wind and solar).

The real concern is the high import dependency linked to a single supplier, i.e. a monopoly. EU competitiveness litigation should in time prevent these gas and oil import monopolies (usually Russia or the Middle East).

Indigenous production was until recently declining, but with the advent of unconventional gas and oil supplies, the rise of nuclear power stations (especially in France, Lithuania, Bulgaria, Slovakia, Belgium and even the UK and Finland), plus the development of wind farms (around the North Sea) and solar power (Spain and Italy) the situation should improve, especially if combined with energy efficiencies.

Figure 2.8 shows EU electricity production in 2007. Note the non-diverse mixes shown from Poland (largely coal), Lithuania and France (largely nuclear) and Sweden (predominantly HEP).

3 Using case studies

Questions
1 Explain how cooperation between EU members can strengthen the EU's energy policies.
2 Study Figure 2.8. Describe and explain the variation in the mixes of electricity generation shown.
3 In your view, should EU energy decision-makers cooperate with Russia, to establish stable working relationships with such a major provider of oil and gas?

Guidance
1 Use http://ec.europa.eu/energy to explore the three facets of the EU Energy Strategy (Sustainability, Competitiveness and Security) and explain how cooperation can support them.
2 Devise a framework, by EU country, providing an overview of amount of electricity produced and the percentage contribution of different energy sources, suggesting reasons in each case (e.g. why does Sweden favour HEP, or why does Poland favour coal?)
3 Include details of recent disputes — the EU was in many ways an innocent bystander. Were the disputes political or economic? Include details of the new pipeline networks.

In conclusion, Part 2 explores the current global consumption patterns, emphasising the overriding dependency on fossil fuels. The issues related to this will be explained further in Parts 3 and 4, giving you an opportunity to evaluate whether there is a need for change and, if so, what the **drivers of change** are.

Part 3

Environmental and socioeconomic issues

Part 3 seeks to evaluate the environmental and socioeconomic externalities of our existing energy supply, which is overwhelmingly dependent on fossil fuels for both primary and secondary use. Figure 3.1 shows the combustion products of the three fossil fuels — not only carbon dioxide and carbon monoxide (both greenhouse gases and key contributors to global warming), but also nitrogen and sulphur dioxides, which are contributors to atmospheric pollution and the formation of acid rain. Atmospheric pollution, illustrated by the Asian brown cloud and graphic 2008 Olympics images of a smog-bound Beijing, is a major problem in the largely coal-fuelled economies of the emerging industrial giants of China and India. It also affects other countries such as the USA and some EU countries, which see coal as a short-term solution to insecurity.

Note how natural gas is by far the cleanest of any of the fossil fuels — it clearly has a key role to play in future energy policies — in particular in the USA and Europe. In addition to fossil fuels, our existing energy supply includes nuclear, hydroelectricity and increasing amounts of other renewables such as wind and solar, all of which make significant contributions to electricity consumption. Table 3.1 evaluates the largely negative externalities caused by all energy sources currently used.

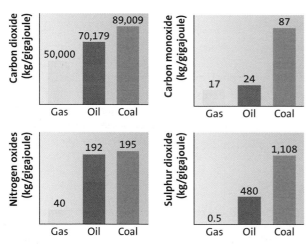

Figure 3.1
Combustion products of fossil fuels (in kg per gigajoule)

Table 3.1 makes the point that there is no such thing as a completely green energy source, although only fossil fuels, especially coal, play a major role in enhancing the impacts of greenhouse gases. Often large-scale schemes that are more economically viable cause more issues, with national interests conflicting with local concerns. The conflicts are particularly complex when energy sources are discovered in environmentally fragile areas or a large-scale pollution event, such as the *Deep Water Horizon* Gulf of Mexico disaster, occurs. *Case study 5* assesses the size of carbon footprints caused by the various sources of electricity generation in the UK and most other developed countries.

Table 3.1 Environmental and socioeconomic impacts of different energy sources

Energy source	Environmental	Socioeconomic
Coal	The most polluting source of energy (greenhouse gases, acid rain and smog); underground mines lead to surface subsidence and toxic waste and water. Opencast pits (the way ahead) scar the landscape, destroying forests and blasting away mountain tops. Overburden buries ecosystems. Whilst legislation requires restoration, the new ecosystems are of low quality.	Mining is dangerous (10,000 deaths per year). Health issues of lung disease for miners and pollution-based diseases for those living near smoke stack industries and coal-fired power stations (scrubbers can be installed). Mine closures lead to derelict landscapes and high rates of deprivation for mining communities (structural unemployment).
Oil	Infrastructure from large oilfields visually pollutes a large area. Oil spills at production sites, along pipelines and tanker routes are ecologically disastrous. Oil/petroleum creates pollution and high levels of greenhouse gases (less than coal). Risk of explosions/fires at tank farms. Ecological issues of exploration in fragile, environmentally sensitive areas.	As exploration goes deep-sea the dangers increase — on average 500 deaths from oil rig accidents per year. Risk of explosions from oil-based installations — a concern for local householders.
Natural gas	Generally seen as the cleanest of the fossil fuels in greenhouse gas terms. Flare-off as a waste product of oilfields causes major environmental problems.	Fears of explosions from LNG terminals and pipelines — especially for local people living in the area.
Nuclear	While nuclear power is clean in greenhouse gas terms, issues include: power plant explosions and accidents that could release/spill radiation into air, land and sea. Toxic radioactive waste and problem of disposal. Terrorist or rogue state use of nuclear fuel for weapons or political leverage. Visual eyesores in attractive coastal locations.	Possible risk of accidents, and increase of cancer (clusters) near nuclear installations. Issue of nuclear power plants decommissioning after 30–40-year cycle leads to mass unemployment where few alternatives exist.
Biomass	The externalities vary with the type used. Collection of fuelwood can lead to deforestation and soil erosion and ultimately desertification. Biomass use of various types can cause atmospheric pollution. Growing biofuels (e.g. palm oil in Indonesia) destroys areas of high biodiversity.	The use of biomass-fuelled cooking stoves is a health hazard, while collecting fuel wood places a huge social burden on women and children as supplies are increasingly scarce. Growth of biofuels contributes to diminishing food supplies: less land for food crops.
Hydroelectricity	Generally perceived as a clean form of energy — largely for electricity generation. Problems include impacts of mega dams, flooding of villages to build reservoirs, trapping of silt leading to erosion, links to earthquakes and methane gas explosions, obstruction of river movement for aquatic life and deterioration in water quality.	Major disruption of villagers' lives on relocation. Reservoirs can be a source of diseases such as cholera or bilharzia. Micro-hydros can lead to a huge improvement in the quality of life of villagers.
Wind	Nimbyism leads to fears of visual and noise pollution, bird and bat kill from turbines and disruption of military flight programmes	Few social impacts, but in places concerns focus on the intermittent nature of the power produced (needs strong, continuous winds).
Solar	Various types but large arrays of cells take up enormous areas of land and are visually polluting in desert areas.	Issues of outsourcing power from poor nations to rich ones.
Tidal	Huge potential ecological impacts on estuaries.	Very large scale — may affect other estuary activities.
Geothermal	Impact of sulphur leads to pollution and infrastructure corrosion.	Abundant hot water on tap adds to quality of life.

A carbon footprint is the total amount of carbon dioxide and other greenhouse gases emitted over the full life cycle of a process or product. It is expressed as grammes of carbon dioxide equivalent per kilowatt hour of generation (g CO_2eq KWh^{-1}). Using the life cycle concept, as shown in Figure 3.2, all electricity generation systems have a carbon footprint because at some point during their construction and/or operation carbon dioxide is emitted.

Fossil-fuel technologies (coal, oil and gas) inevitably have the largest carbon footprint because they burn these fuels during operation. Non fossil fuel-based technologies, such as wind, solar (photovoltaic), hydro, biomass, wave/tidal and nuclear are often referred to as **low-carbon technologies** or even **carbon neutral** as they do not emit carbon dioxide during their operation. However, carbon dioxide emissions do arise in other phases of their lifecycle, such as during extraction, construction and maintenance, and in decommissioning in the case of nuclear power stations.

Figure 3.2
Life-cycle carbon dioxide emissions for electricity generation technologies

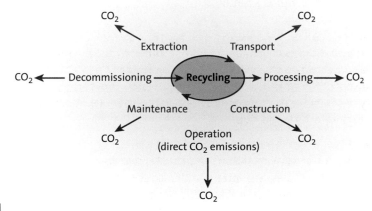

Coal

Coal-burning power systems have by far the largest carbon footprint. Conventional coal combustion systems result in emissions in the order of 1,000 g CO_2eq KWh^{-1}, although this can be lowered to 800 g CO_2eq KWh^{-1} using newer gasification plants (so-called clean coal plants). Future developments, such as carbon capture and storage (CCS), or co-firing with biomass have the potential to reduce the carbon footprint of coal-fired electricity stations still further. Trials of co-firing wood with coal are taking place at coal-fired power stations in the UK, for example at Drax. Biomass energy crops are currently more expensive than coal, but by co-firing biomass power station operators can earn Renewable Obligation Certificates, which subsidise co-firing and make it economically viable. E.ON built the UK's largest dedicated biomass plant (44 MW) at Lockerbie, Scotland in 2007.

Technological improvements for cleaner coal could increase the energy efficiency of existing coal-fired plants from current levels of 35% to over 50% conversion into electricity. CCS could potentially avoid 90% of carbon dioxide emissions.

Oil and gas

Oil accounts for only a very small proportion (less than 5%) of electricity generated in the UK — it is primarily used as a backup fuel to cover peak electricity demands — but the average carbon footprint is high at 650 g CO_2eq KWh^{-1}.

The dash for gas has led to over 40% of UK electricity being generated by gas, with the footprint at 500 g CO_2eq KWh^{-1}, half that of coal because gas has a lower carbon content and higher calorific value. Gas plants could co-fire biomass to reduce carbon emissions still further in the future.

Low-carbon technologies

In contrast to fossil-fuelled power generation, the common feature of renewable and nuclear energy systems is that emissions of greenhouse gases and other atmospheric pollutants are indirect and therefore low overall.

Figure 3.3 shows the range of carbon footprints and compares low-carbon with fossil fuel technologies (current and future).

Biomass energy is obtained from organic matter, either directly from dedicated energy crops like willow and *Miscanthus*, or indirectly via biomass conversion from wood chips. Biomass is classified as 'carbon neutral' because the carbon dioxide released by burning is equivalent to that absorbed by the plants during growth. But this calculation excludes emissions arising from fertiliser production, harvesting, drying and transportation of biomass. The range of carbon footprints for biomass is related to the type of organic matter and the way it is burned — low-density *Miscanthus* contrasts with high-density wood chips.

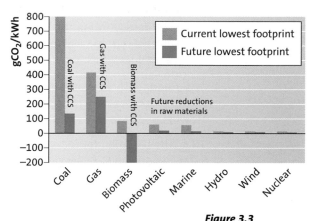

Figure 3.3
Current and future carbon footprints (UK, European, US and Australian power plants)

Photovoltaics (PVs)

Solar cells are made of crystalline silicon that is extracted from quartz sand at high temperatures. The process is energy intensive and adds to the carbon footprint. New technology has been developed using thin-film silicon, the production of which is less energy intensive, thus decreasing the footprint. Note that in southern Europe, lifecycle carbon dioxide emissions are lower because with more sunlight, operating hours are greater and energy output higher.

Waves and tides

These are emerging technologies. Most carbon dioxide is emitted during the manufacture of the structural materials used to make the turbines.

Hydroelectricity

There are two main types of hydroelectric scheme:
- Storage schemes require dams and so have a high construction footprint. Once in operation these schemes emit very little carbon dioxide, although some methane emissions do arise due to decomposition of flooded vegetation.
- Run-of-river schemes do not give rise to significant footprints.

Wind

Electricity generated from wind energy has one of the lowest carbon footprints. In the future, improvements in the manufacturing of turbines and construction phases could lower this still further. Emissions generated during operation include routine maintenance inspection trips. Note that this is marginally more intensive for offshore wind turbines.

Nuclear

Nuclear power generation has a relatively small carbon footprint, since there is no combustion (heat is generated by fission of uranium or plutonium). Most emissions occur either at the mining and enrichment stage or during decommissioning and construction of waste storage facilities. If global nuclear generation capacity increases, higher-grade uranium deposits could be depleted so more energy would need to be used, thereby increasing the carbon footprint. Manufacturing energy from recycling spent fuel rods has a higher carbon footprint, but still only similar to other low-carbon technologies.

Carbon footprints could be further reduced for all electricity generation technologies if the processes were themselves fuelled by low-carbon technologies, for example using wind power to make the steel for wind turbines.

As the situation for UK electricity generation stands today, the overwhelming drivers of change are fulfilling carbon dioxide emission targets and managing energy security to overcome the potential 'peaking' in North Sea oil and gas supplies. The big issue for the UK is how best to achieve these twin aims and manage the gap between decommissioning ageing nuclear and coal-powered stations, and moving to low-carbon future technologies (see Parts 5 and 6).

When looking at the current overwhelming reliance on fossil fuels, it becomes clear that in terms of non-transport use coal is going to continue to play a vital role (see Figure 1 page 6), yet it is coal which is the overwhelming focus for concerns about existing fuels.

Coal: the comeback kid?

This section explores the controversies surrounding coal as a future mainstay of global energy production.

Coal currently provides around 30% of the world's electricity, although the percentage is much higher in China and India, where the growth in demand could lead to the burning of 7 billion tonnes annually by as early as 2030. The potential for the use of coal depends on a number of factors, not least the prices of alternatives. Currently, in terms of direct cost, it is a relatively cheap fuel, although prices are rising. For this reason, coal consumption has risen 50% since 2000 largely driven by Chinese demand — China is now a net importer. However, increasing freight costs, higher mining costs (especially from deep mines) and ultimately the costs of cleaning up coal make the economic equation less favourable.

While traditionally coal has been seen as a widely distributed and abundant reserve with up to 200 years before 'peak' is reached (IEA), the reality may be that global coal output will peak as early as 2025 and from then decline. Much of the easily mined high-quality hard coal has been used and the economically recoverable reserves are declining rapidly. There are major concerns over the reliability of the data on reserves, with Russia and China, two of the world's biggest producers, failing to provide up-to-date figures. The latest official statistics from the World

Energy Council, which put global coal reserves at a staggering 847 billion tonnes (compared with 2007 production of just under 6 billion tonnes), have to be treated with caution.

Environmental and social impacts of coal mining

By far the most economic means of mining coal is in huge, open-cast pits. Environmentally this is extremely damaging.

Mountain-top removal is an extremely destructive method of open-cast mining. The hills and hollows are stripped of trees and vegetation and streams are buried under tonnes of rock and debris (see Figure 3.4). One such scheme near the village of Bob White, West Virginia, was successfully challenged by environmentalists in the courts. In south Wales at Ffos y Fran near Merthyr Tydfil, there are plans to excavate 10.8 million tonnes from a hilltop, which will take nearly 20 years to dig out and then refill. This is just one of around 22 new open-cast schemes currently planned for the UK and scheduled to produce 55 million tonnes of coal as part of a series of stop-gap measures to fill the electricity shortfall caused by decommissioning of older coal-fired power stations and several large nuclear plants.

Figure 3.4
(a) Mountain-top removal in West Virginia. (b) The location of Bob White

Climate change protestors focus on the building of any new, large, coal-fired power station, whether it be in Kent, Texas, Virginia or even in China.

There are possible plans for 150 new coal-fired power stations in the USA. In Texas, for example, the energy company TXU wants to add 11 coal units to its existing mix of power stations, at a cost of nearly $1 billion a piece. This has led to a firestorm of opposition. One environmental group calculated that the new plants would generate more carbon dioxide each year than all the emissions from Denmark.

Figure 3.5
Environmental protesters close to the site of the proposed Kingsnorth Power Station in Kent (now not to be built)

TXU sees this as just a beginning — it argues that coal is abundant in the USA, cheaper currently than natural gas, more reliable than wind power and easier to get planning permission for than nuclear proposals. Even if gas prices were to fall and coal were to be heavily taxed, the proposals seem economically robust, unless renewables become really cheap in the next 20 years.

TXU says it will deploy state-of-the-art technology, such as scrubbers, to trap pollutants and that these new power stations are replacing old ones, so emissions of oxides of sulphur and nitrogen will drop by 20%. Environmental protesters argue that TXU should employ integrated gasification combined cycle (IGCC) technology, which turns coal into synthetic gas so it can be sequestered underground — currently an experimental process — or that the power stations should all be CCS ready.

Equally controversial are plans to build schemes in Germany (25), South Africa (a massive World Bank-funded project at Medapu) and Australia. Even in power-hungry China and India there has been some resistance to new coal-fired power stations — all part of the growing coordinated response from the Climate Action Network. The Chinese Environmental Protection Agency actually closed 19 new coal-fired power stations because they had failed to undergo environmental impact assessments, although, to put this in perspective, China has been building around 70 new coal-fired power stations each year and plans over 500 more).

Aside from environmental concerns, there are also social externalities to consider. Open-cast mines generally have much lower social costs on miners' health, as *Case study 6* shows.

Contemporary Case Studies

COAL'S TOLL: IT'S DIRTY AND DANGEROUS — Case study 6

One major social externality of coal is the fact that deep mining is both a dangerous and dirty business. Every year, headlines reinforce this: 'No survivors in New Zealand mine', 'Death toll rises to 104 after a huge gas blast in Heilongjiang Province in NE China' or 'Lax operators keep unsafe mines running — 29 killed at the Upper Big Branch mine in West Virginia USA'. And then there are the high levels of morbidity in traditional deep mining areas with ill health from dust-related diseases such as pneumoconiosis — the men frequently die before their complex compensation claims are settled.

China

As the largest producer and consumer of coal, and with a staggering 5 million coal miners, China, until recently, was experiencing some 7,000–10,000 mining-related deaths per year. (The USA averages 40 deaths a year.) There was no such thing as a safe mine, with low levels of mechanisation, few mining engineers and poorly trained workers. China relies on deep-shaft mining, and safety equipment is still rudimentary (2009). Some 600,000 Chinese miners are thought to suffer from black lung disease.

China's booming economy has led to a voracious demand for power — around two new thermal power stations are opened each week. Provincial supplies of coal are often precarious with coal arriving by lorry trains on China's congested roads, and power stations finding themselves sometimes with only hours of coal supply left. Entrepreneurs have opened coal mines across the country, with little regard for working conditions and safety.

In recent years the government has instituted a safety drive that has cut the deaths since 2004 to around 3,000 per year, primarily by the forced acquisition or closure of small, private, often illegal mines. Investigations into previous mining accidents suggest that such was the pressure to achieve production, managers juggled risk with reward and refused to clear the workers from pits, even when gas monitors indicated problems and evacuation alarms sounded.

On the plus side, mining offers relatively high wages to rural dwellers, for whom little alternative work is available. A compensation system for mine fatalities has also been introduced by the Chinese government.

Other countries

The **USA** has the world's biggest coal reserves. As coal prices rise, abandoned mines are reopened, and new ones developed. Over half of US coal is mined in the northwest, in Utah and Wyoming, largely from safer, huge, open-cast pits, which are strip mined. The second most productive region is Appalachia (Pennsylvania/Kentucky and West Virginia) where new deep mines are opening, leading to a shortage of miners. It is here that the deaths usually occur, with some recent major accidents (Upper Big Branch, 2010).

Russia: Russian miners for decades enjoyed 'hero status' and were in relative terms well paid. Over the last 20 years, many of the dangerous deep mines have shut, and 70% of the output comes from open casting, which is both less dangerous and employs fewer people. Such deaths as there are come from methane explosions in badly maintained and poorly ventilated deep mines.

South Africa produces some of the cheapest coal in the world largely from open-cast or shallow-drift mines in Mpumalanga Province. For this reason, it is a major coal exporter. One reason for the cheap costs is the low wages of the almost exclusively black workforce. Overall the safety record is good, and many of the mining companies supply their workers with antiretroviral drugs to combat the high incidence of HIV/AIDS.

Poland's upper Silesia, centred on Katowice, is a typical heavily industrialised coal mining area. When Poland became a member of the EU, it had to cut back on its subsidies to its coal industry, so many mines closed or became privatised and up to 350,000 miners lost their jobs (leading to structural unemployment rates as high as 20%). As with any deep-mining area, gas explosions did occur, but overall the safety record was good for a deep-mining area. Nearly all of Poland's electricity is generated using coal.

The conclusion would seem to be that deep mining can be very dangerous, if there is no strong legal framework for health and safety standards and mine safety is not monitored. Many economists argue that the social costs incurred should be factored into the price of coal.

Using case studies

Question

To what extent do you agree with the statement that the social costs of coal mining are even greater than the environmental costs?

Guidance

Social costs are especially great for deep mining and could be managed largely by improved health and safety regulations. Direct environmental costs are higher for open casting but indirect costs from power station emissions are of a global scale. You need to evaluate the scale of the problems and also their difficulty to solve.

While few will argue about coal's status as a fuel of convenience, its long-term future and opportunity to make a comeback lies with convincing a world battling the impacts of climate change that clean coal is technologically feasible. As coal's main use is in steel works and power stations, this new technology will be vital.

Figure 3.6 summarises the range of coal supplies and their main uses.

*Figure 3.6
Types of coal and their uses*

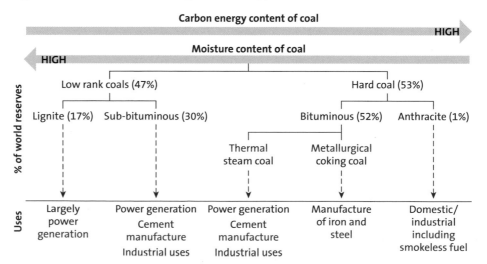

Cleaning up coal

The important question therefore is can we clean up coal and, if so, how quickly? The low efficiency of current power stations is one reason why coal is so bad for the environment, as the vast majority of older coal-fired plants operate by burning coal to produce steam. Carbon dioxide emissions can be lowered by getting more energy from the carbon that we use.

The immediate solution is to introduce **clean coal technology** either by using PC (pulverised coal), which increases efficiency by up to 15%, or **gasification**, which is where coal is first burned to create a gas, which can then be burned in a gas turbine (IGCC). There are already several commercial IGCC plants in the world (USA, Spain and Netherlands). **Underground gasification**, currently being trialled in China and Australia, is the latest development. This involves reacting entire coal seams with oxygen and steam in situ, and pumping out the mix of methane and hydrogen this creates to be burnt in generators.

Carbon capture and sequestration

In the long-term, carbon capture and sequestration (CCS) is seen as the 'magic bullet solution', with pilot projects being developed around the world, including the UK. CCS involves capturing the carbon dioxide released from burning fossil fuels, transporting it to a storage site, and pumping it underground so that it is not released to the atmosphere.

As the technology develops, CCS would allow power plants that run on fossil fuels — particularly coal-fired power stations — to continue operating with vastly reduced emissions (maybe by up to 90%). Many of the components that are required for CCS are available today, but the end-to-end process has yet to be demonstrated at full scale.

There are two main barriers to the deployment of large-scale CCS:

- **Cost** — capturing the carbon dioxide reduces the efficiency of power plants by 20–25%, and transporting it to storage sites through new pipeline infrastructure will be expensive.
- Uncertainty about the **reliability** of storage — it is not clear how long carbon dioxide will stay underground; it will be important to demonstrate to the public that storage is safe and reliable. Already in Germany huge public protest at a demonstration plant has halted the storage process.

Current research activities are focused on all three stages of the process:

Capture

Three alternative options for capturing carbon dioxide from power plants are under development:

- **pre-combustion capture** — gasification of the fuel into hydrogen and carbon dioxide and then using the hydrogen to generate power
- **oxy-fuel combustion** — burning the fuel in pure oxygen to create a carbon dioxide stream that can be captured
- **post-combustion capture** — capturing the carbon dioxide emitted in combustion through chemical absorption

Transport

Transporting the carbon dioxide to the storage site will involve building new infrastructure:

- **pipelines** to transport the carbon dioxide to the storage sites, both on- and offshore
- **ships** to transport the carbon dioxide to offshore storage sites if no pipelines exist

Storage

There are a number of options for storing carbon dioxide underground:

- **saline aquifers** (rock formations that hold water) have the largest potential capacity
- **depleted oil and gas fields** have a potential capacity several times smaller than aquifers
- **enhanced oil recovery sites** — smaller-capacity but likely to be developed first because the carbon dioxide injection improves the economics of the oilfield; however, this benefit only lasts until the oilfield is depleted, and some of the carbon dioxide will be extracted with the oil

Using CCS

A number of pilot plants are in operation today to test CCS, and large-scale demonstrators are planned in Europe, North America, Australia and China over the next 5 years. The UK plans to pipe carbon dioxide emissions from the Killingholme power station to the depleted gas fields in the North Sea.

At its best CCS could prevent 90% of the carbon dioxide emitted by power stations from getting into the atmosphere.

With over 1,000 new coal-fired power stations planned (70% of these in China and India) it is vital that action is taken to clean up coal and that a low-carbon energy revolution begins. While the World Coal Institute argues that CCS has already been deployed at a number of sites around the world (but not yet at power stations), Greenpeace argues that CCS is an unproven technology and will not be commercially available until 2030 — by which time it may be too late. Households and businesses may be largely responsible for making the required investments, but it is governments that hold the key to changing the energy mix via policies and regulatory frameworks.

Bridging the gap

The real issue is how to bridge the gap between the existing energy supply mix and the renewable fuels of the future. This leaves us with five key questions:

- Can we clean up coal soon enough?
- Is nuclear power the most suitable stopgap option?
- Will unconventional supplies of gas be the answer?
- Will renewable technologies become economically viable soon enough?
- Are energy efficiencies the real solution?

Conservation of energy can lead to radical changes in lifestyle, but it may be an essential component if we have to accept higher energy prices.

Case study 7 shows how the issues of existing energy supply have spilled over to a possible crisis in the UK's electricity supply.

ELECTRICITY IN THE UK: HEADING TOWARDS A DARK FUTURE

Figure 3.7 summarises Britain's shortfall of electricity supply. The energy regulator has warned of power blackouts and the possible need for the renationalisation of electricity supply as the six major private electricity generating companies have failed to invest. The UK's electricity security problem in 2010 was caused by under investment, with political opposition and planning hold ups affecting nuclear, coal and even low-carbon alternatives such as wind (onshore and offshore). The shortfall is exacerbated by plant limitations, such as ageing, coal-fired power stations without scrubbers, and the decommissioning of large, nearly 40-year-old, nuclear plants (see Part 5). From 2010 to 2020, 42 plant retirements are planned — more than in any other EU country. There will be a cumulative drop of 34% in electricity generating capacity by 2020 and it is predicted that demand for electricity will increase.

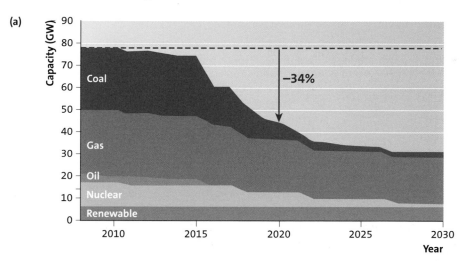

(a)

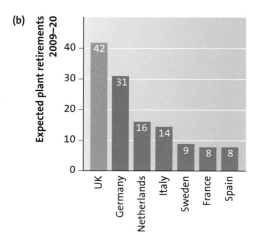

(b)

Investment programme
Total UK energy investment required by 2025: £233.5bn

includes:

Wind and renewables	£112.5bn
Nuclear power stations	£38.4bn
National Grid upgrades	£24bn
Smart meters	£13.4bn

Figure 3.7
The UK's electricity supplies: (a) the shortfall in power (cumulative existing power station requirements); (b) expected plant retirements 2009–20 and the UK's investment programme until 2025

Questions

1 Explain why UK electricity generation capacity would seem to be heading towards a dark future.
2 Assess the suggested investment programme planned by 2025. Do you agree with the proposals? What further investment would you make?

Guidance

1 Analyse Figure 3.7 in detail. Identify the reasons for the electricity gap. Try to research some named examples of the planned retirements.
2 Work through the options shown.

Nuclear power stations — low carbon — have other issues — also take around 15 years to build

Renewables — high cost of offshore wind

Smart grids clearly a sound infrastructure upgrade — smart meters would clearly cut down consumption

Other investments could include:
- clean coal
- shale gas (Morecambe Bay and Ribble Valley already successfully explored) — gas power stations clean and efficient — current world price and supplies encouraging news — also quick to build to fill the gap

Running out of oil and gas?

Looking into the future

We are addicted to oil; it is the lifeblood fuelling our economies and allowing comfortable, often global, lifestyles for growing numbers of peoples. It increasingly dominates global politics and nations go to war for it. Oil and gas are fundamental building blocks of the modern world, but there is now serious discussion about when this oil age might end, and how the world might face up to this.

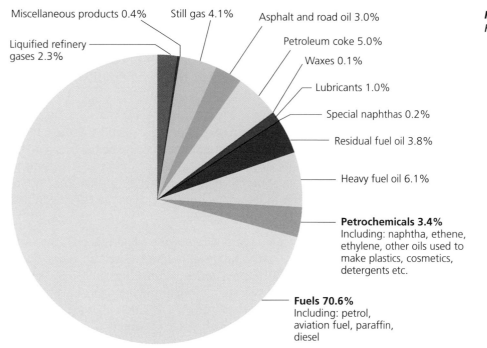

Miscellaneous products 0.4% Still gas 4.1% Asphalt and road oil 3.0%

Liquified refinery gases 2.3%

Petroleum coke 5.0%

Waxes 0.1%

Lubricants 1.0%

Special naphthas 0.2%

Residual fuel oil 3.8%

Heavy fuel oil 6.1%

Petrochemicals 3.4%
Including: naphtha, ethene, ethylene, other oils used to make plastics, cosmetics, detergents etc.

Fuels 70.6%
Including: petrol, aviation fuel, paraffin, diesel

Figure 4.1
How oil is used

Figure 4.1 shows a breakdown of how oil is used. While 70% is used as fuel, largely for transport, the pre-tax value of this is just about the same as that of the 3.4% used to provide petrochemical products: plastics, cosmetics, detergents, pharmaceuticals and paint — all essential for 'modern living'. These valuable by-products are the hardest to replace. Although research is being done in bio-refineries, and in using

www.g-wiz.org.uk/information-electric-car.html

Figure 4.2
*A 'G-Wiz' electric car.
Transport problems
may be more
easily solved using
alternative fuels*

biomass-based feedstocks, this in itself is a very energy-intensive process. As for transport, the problems are possibly easier to solve by using alternative fuels such as hydrogen, biofuels or even gas, in combination with developing more fuel-efficient vehicles (Figure 4.2).

While nearly everyone can see the need to reduce the world's overdependence on oil the solutions require some radical changes in attitudes and lifestyles.

Supply issues

The first threats to supplies, other than in wartime, occurred during the OPEC-led embargo in 1973, when it first became apparent that oil could be used as a political weapon. In the early 1960s the seven major Western energy TNCs controlled exploration and production, but the formation of OPEC broke the Western stranglehold. The 1973 oil embargo showed countries such as the USA a glimpse of a world with 'rationed' or very limited oil supplies.

The power of oil as a political weapon is amplified by the fact crude oil is the world's most actively traded commodity in the financial markets. Any bad news, such as BP's *Deepwater Horizon* spill in 2010, or a series of unsuccessful oil strikes, has repercussions around the world, causing shares to plummet and oil prices to 'spike'. Stock markets gambling on oil shortages caused an escalation in crude oil prices to nearly $150 in late 2008 when there was no shortage of supply.

As stated in Part 2, it is the concentration of supplies (70% of oil, 65% of natural gas) in the **strategic ellipsis** that stretches from the Middle East via the Caspian Sea to northwest Siberia, which has reinforced their strategic value as geopolitical weapons. The fear of oil insecurity has shaped US and EU political decisions — some say that it was an important factor in driving the war with Iraq. This lack of security is further exacerbated by choke or pinch points in the global supply

Figure 4.3
Events such as the BP Deepwater Horizon rig explosion and oil spill in the Gulf of Mexico have repercussions around the world

US Coastguard

networks of pipelines and supertankers. Knocking out just one of these choke points could have huge short-term consequences in upsetting the supply/demand equilibrium.

A number of factors can cause disruption along pathways followed by pipelines and supertankers.

■ Pirates cause problems south of the Straits of Hormuz and in the Straits of Malacca.

Figure 4.4
Danger zones at pinch/choke points

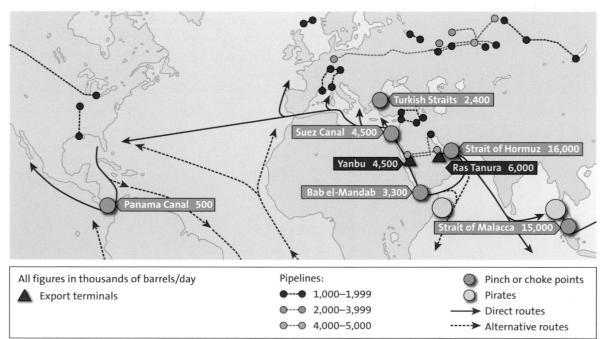

Turkish Straits 2,400		
Suez Canal 4,500		Strait of Hormuz 16,000
Yanbu 4,500		Ras Tanura 6,000
Bab el-Mandab 3,300		
Panama Canal 500		Strait of Malacca 15,000

All figures in thousands of barrels/day
▲ Export terminals

Pipelines:
●--● 1,000–1,999
●--● 2,000–3,999
●--● 4,000–5,000

● Pinch or choke points
○ Pirates
→ Direct routes
-----▶ Alternative routes

- Accidental explosions on oil rigs, along pipelines or at oil refineries (e.g. Abqaiq in Saudi Arabia), or spillages (e.g. along the Trans-Alaska pipeline) can interrupt supplies for several months.
- These accidents have the potential to be replicated by terrorists, for example in Iraq (see *Case study 10*) or Nigeria (see *Case study 12*). Such attacks could be especially effective at choke points (for maximum impact). Hijackers could even commandeer a LNG super tanker and load it with explosives to make a floating bomb. This could generate a burning oil slick which could render a busy route impassable for several years.
- Weather events such as hurricanes in the Gulf of Mexico can lead to a stoppage in oil pumping for up to a fortnight.
- Conflicts such as the Gulf War or those between Iran and Iraq, where oil capacity was deliberately destroyed, can disrupt production for years.
- Saudi production is especially important because of its dominance. It would only take one 9/11-style event to cripple a huge export terminal such as Ras Tanura or Yanbu, leading to a long-term shortage of supplies.

There is therefore a growing realisation that we could see the world in economic meltdown if there was a significant glitch in the availability of oil supplies. Until the 2008–09 recession the balance between production and consumption was very tight in the short term. The question is when, not whether, peak oil will occur, leading to longer-term problems.

The controversy of peak oil

Peak oil is usually defined as the point at which the rate of consumption of oil reserves is not matched by oil coming on stream from new discoveries, i.e. a production peak. As the world's population is still increasing (albeit at a decreasing rate) and the demand for energy is rising even faster, largely because of demand from emerging superpowers, consumption is growing. It took nearly 125 years to use the first trillion barrels, yet the next trillion will be used in less than 30 years.

The real problem is that we do not know exactly when peak oil will occur. Pessimists such as the Association for the Study of Peak Oil say it has already occurred, whereas optimists (including some oil companies) think it may occur as late as 2050. The International Energy Agency (IEA) has now revised its estimate upwards to 2020 — as a more-or-less firm prediction of when oil supplies will cease to flow, based on an analysis of production trends in 800 individual oilfields in 2008.

Why is it so hard to forecast?

So why this discrepancy in such an important forecast? One problem is that there is confusion and controversy about how large oil reserves really are. The BP report (2010) estimates that crude oil reserves will last for at least 50 years, with a sloping plateau of production decline, so much depends on the success of new prospecting.

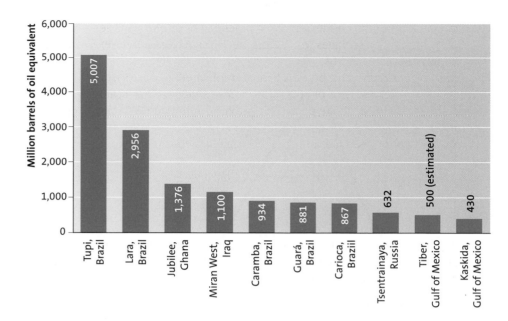

Figure 4.5
Oil and gas fields discovered since 2006 (million barrels of oil equivalent)

As Figure 4.5 shows, as a result of increased reliability of geological data, there have been some large recent finds, especially in Brazil (see *Case study 8*). In addition, the price per barrel determines how oil production compares cost-wise with other fuel resources such as renewables, and whether the use of unconventional oil from tar sands and oil shales becomes economically viable.

Many of the new discoveries are not easily accessed, being found in:
■ deep seas (e.g. offshore of Brazil, Mexico, Libya), which are technologically complex to drill
■ environmentally fragile areas such as the Arctic, where costs are high
■ developing nations, where there are skill shortages (e.g. Angola)
■ former war zones (e.g. Iraq)

All these sources could delay the onset of peak oil — but is the cost acceptable?

The general feeling from experts in the oil industry is that there is no new giant oilfield to be discovered, except perhaps in Antarctica (the world's last pristine wilderness).

As oil is such a strategic resource, certain energy TNCs and producing countries are reluctant to declare their true reserves and may underestimate or exaggerate them. For OPEC members these declarations can determine their quotas and rates of production, and therefore control their income, while other countries do not accurately survey their reserves. TNCs are reluctant to share such information for commercial reasons.

Figure 4.6 uses a generalised logistic curve (usually known as a bell curve) to illustrate peak oil. The concept was first developed by Hubbert. Parameters of new discoveries, production and consumption rate and price per barrel can be adjusted to control the slope, height and timing of the peak.

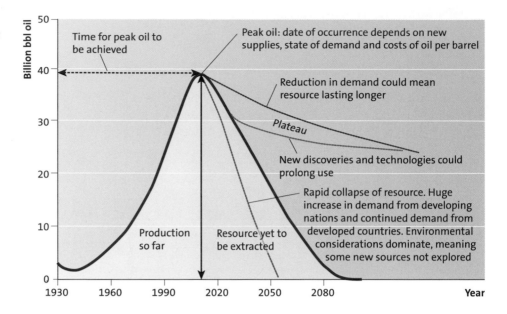

Figure 4.6
A model for peak oil using a generalised logistic curve

Chart labels:
- Billion bbl oil (y-axis)
- Year (x-axis)
- Time for peak oil to be achieved
- Peak oil: date of occurrence depends on new supplies, state of demand and costs of oil per barrel
- Reduction in demand could mean resource lasting longer
- Plateau
- New discoveries and technologies could prolong use
- Rapid collapse of resource. Huge increase in demand from developing nations and continued demand from developed countries. Environmental considerations dominate, meaning some new sources not explored
- Production so far
- Resource yet to be extracted

Are we running out?

The case for early global oil depletion is supported by the following:

- Oil production has declined in countries such as the USA, which possibly reached peak oil in the 1980s, or in the North Sea, which may have reached its peak in 2008 (although in both cases new finds have occurred).
- Fewer giant oilfields are being discovered and about 80% of those developed before 2005 show production declining at an annual rate of about 6%.
- Reported reserves may be inflated to begin with (possibly by oil companies).
- Emerging nations such as China and India have increased their demand so much that this has placed great pressure on future global supplies.

The counter arguments against imminent peak oil are as follows:

- The world has never run out of any globally significant non-renewable resource as increased prices tend to dampen down demand and stimulate increased technological efficiency and the development of substitute materials.
- Recent discoveries and production in global oil and especially gas (see *Case study 2*) have not declined as predicted, with unconventional supplies contributing to a brighter future.
- Hubbert's forecast for the USA has proved far too pessimistic, as the country has developed new indigenous resources (see *Case study 2*) even though many are unconventional and potentially dirtier.
- Higher prices will stimulate further exploration, with improved technology leading to more of a plateau than a peak.

What do you think about peak oil? Search the web for 'peak oil' to explore the concept further.

Players in the oil and gas industry

A number of players will have a major influence on the future of oil. Their roles and likely impacts on oil supplies are summarised in Table 4.1. Problems arise where key players conflict with each other, for example in western Siberia.

Table 4.1
Players in the oil and gas industry

Player	Role	Degree of influence
OPEC Organisation of oil producing and exporting countries — a cartel	Twelve members protect interests, stabilise prices by controlling production, and eliminate harmful price fluctuations, so providing an efficient flow to consumers.	**Very high**: with peak oil more imminent the fact that members control 66% of world reserves and 40% of production is of key importance.
Governments In oil-producing countries that are not in OPEC	As oil is discovered in non-OPEC countries such as Norway, and within South America and Africa, government influences increase. Some, such as Russia, totally control foreign TNCs. Others are corrupt and have a bad influence on oil developments.	**High**: some governments exercise sound stewardship and investment strategies, e.g. Norway. Others lack competence and expertise. Together these countries provide the majority of world production. Governments can force TNCs to share developments, e.g. BP and Shell in eastern Siberia.
Gazprom and other state-owned oil companies	Gazprom is a state-owned (52% ownership) Russian TNC. Other state-owned oil companies include Saudi Aramco, NIOC (Iran), Petro China, Pemex (Mexico), NNPC (Nigeria).	**Very high** in Russia, where Gazprom decides where TNCs can invest. Most state-owned companies spearhead exploration and production.
TNCs	Major TNCs include Exxon Mobil, BP, Royal Dutch Shell, Total and Chevron. Banks finance operations. Energy companies develop electricity generation.	The need to make a profit for shareholders pushes them to risky locations. **High** in terms of expertise but ultimately they can only operate within host countries' rules. Powerful because of their diversity and presence in many countries.
Consumers	Essentially 7 billion people as decision makers, e.g. in deciding whether to support alternatives such as nuclear. More likely to be influential in developed countries as investors via shares in TNCs.	**Moderate** as they have influence over preferred energy mix and decisions concerning future scenarios, e.g. picking green options at higher cost, or organising anti-nuclear protests.
Environmentalists	Some such as Greenpeace are worldwide campaign groups; others are local groups, e.g. protestors in Ogoni province Nigeria, or climate-change action groups at the Kingsnorth power station site.	**Moderate to high**, especially globally networked climate-change campaigns against coal, or local protests against environmentally sensitive proposals (Nimbyism). Increasingly effective as becoming more professional.
Scientists, prospectors, technologists	Advise on methods of production, influence resource exploration; also develop new technologies, such as nanotechnology to revive old oilfields.	Not powerful, but have a crucial role, e.g. sealing the leak in the *Deepwater Horizon* rig.
Local people	Resource-based development can change their lives, so they often fight these developments, e.g. shale gas.	Usually **low** — exploration happens round them, but occasionally their influence is high when they win a campaign.

Pushing back the frontiers: new oil

The world consumes 85 million barrels of oil every day (2010) — enough to fill over 5,000 Olympic swimming pools. By 2030 this could reach 105 million barrels per day in spite of the predicted supply crunch (IEA).

Alternatives for both transport and petrochemical products are being actively developed, but oil is convenient. As prices per barrel rise, this will fuel an effort to exploit 'exotic' supplies, revisit older fields to squeeze more oil out of them, and invest millions in unconventional sources. These include:

- tar sands — a mixture of sand/clay and bitumen, forming a viscous, black sticky petroleum-based deposit
- oil shale — a sedimentary rock containing kerogen (a precursor to petroleum)
- synthetic liquid fuels from coal and gas

The last option is heavily dependent on the availability of reserves in these two fossil fuels. As we saw on page 17 the situation is currently more promising for natural gas as a result of the potential of shale gas. For coal (see page 42) expert opinion now suggests that the reserves situation is less promising.

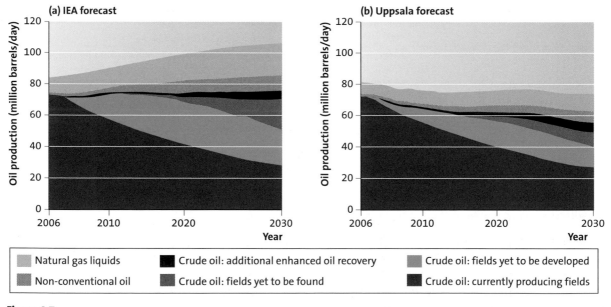

Figure 4.7
'Trouble ahead'— forecasts for global oil production (million barrels per day): (a) IEA forecast; (b) Uppsala forecast

Figure 4.7 summarises the future situation from two different perspectives. According to the IEA, in terms of geological abundance the sources shown look more than sufficient to meet rising global demand, as together they could raise the supply to about 90 billion barrels per day. But this is only one scenario — the Uppsala scenario shows a far more gloomy future.

Figure 4.7 is headed 'trouble ahead' as it emphasises how the easy-to-produce oil is running out. There are two big issues to face. As you will find out, all these new sources are either dirtier/more environmentally damaging or technologically more complex to extract and therefore will be much more expensive to exploit.

Figure 4.8 summarises the situation in terms of availability and cost. Oil exploitation is a long-term investment, and the recent fluctuations in prices — $150 a barrel in 2009, down to $70 in 2010 and back to $100 in 2011 — are difficult for oil companies to manage.

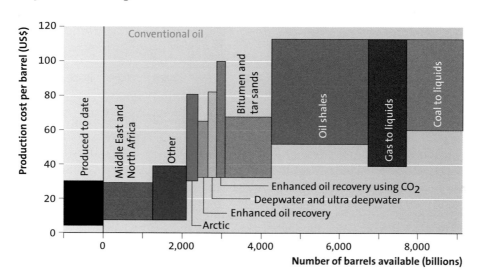

Figure 4.8
Oil cost ranges by source

Extreme oil: scraping the bottom of the barrel?

Enhanced oil recovery

The first approach is to revisit existing oilfields and attempt to extract the residual oil. This is known as enhanced oil recovery and takes place after all the free-flowing oil has been extracted by conventional pumping. A conservative estimate from IEA suggests that some 360 billion barrels remain in old US fields alone. The oil is found in small porous mini traps away from the main oil trap or clinging to grains of sedimentary rock (likened to greasy residue on a frying pan).

Geologists map the residue by seismic imaging — traditionally oil companies flush water through the oilfields to try to recover stubborn residues (up to 30%). The oil companies also add detergent which frees the oil from the surface of individual grains and forms an emulsion which can be recovered and broken down to extract the oil (leading to up to 60% recovery). In the future, nanotechnology will make these processes more efficient still.

Extreme supplies

The second approach is to develop technology to explore further afield, for example the ocean depths or areas with extreme weather conditions.

Figure 4.9
*Maximum
operational depth
(km) of offshore
oilfields in first
operating year*

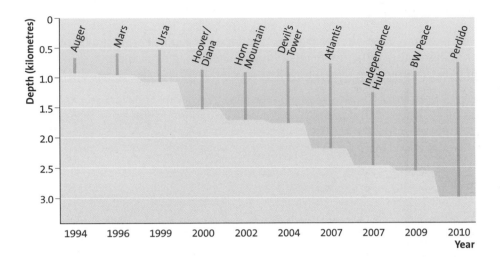

Figure 4.9 shows how improved technology has made offshore drilling feasible to a depth of 3 km, thus allowing exploitation of offshore fields in Mexico, Brazil and Libya. Some of the discoveries, such as the new Tiber field in US waters in the Gulf of Mexico and the new Brazilian fields (see *Case study 8*) are very significant finds (as large as the new Iranian fields) but not as large as the current giant Saudi fields.

The technological expertise required is staggering — for example, to discover oil in the Tiber field the drill had to reach 9.4 km below the sea bed and the exploration oil rig stands in 1.2 km of water. A combination of improved ways of locating oil (the exploration costs of a well that does not strike oil may be as much as $200 million) and improved drilling and platform technology has made these new frontiers possible. However, as the BP *Deepwater Horizon* disaster showed, there are increased dangers of both explosion and spills. The magnitude of the 2010 disaster led to worldwide concerns over deep-sea oil exploration, and in the USA to a moratorium on drilling to review safety considerations.

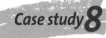 **DEEP-WATER OFFSHORE OIL IN BRAZIL**

In 2001 Brazil discovered the Tupi oilfield and in 2008 the Carioca oilfield, both in deep water some 300 km offshore from the coastal city of Rio de Janeiro. The oilfields could yield at least 20 billion barrels (2009 estimates) of light, high-quality crude and could propel Brazil into the major oil league, with reserves pushing it into the top 10. Ironically, Brazil is currently the world's leading producer of ethanol (largely from sugar cane) — an industry developed because of a lack of indigenous sources of oil and gas. Petrobras, the state-owned Brazilian oil company, has spent more than $8 billion in 'signing up' more exploration rigs for further deep-water exploration in the hope (unlikely) of Brazil becoming Latin America's Saudi Arabia.

As well as deep-water exploitation, oil companies are exploring areas that are environmentally contentious, or politically and economically difficult. *Case study 9* looks at the issues faced in Greenland and *Case study 10* explores the problems in redeveloping Iraq's oil industry after many years of war.

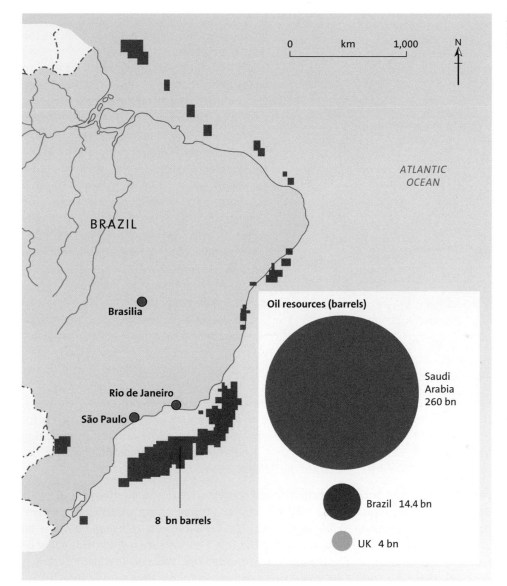

Figure 4.10
Location of the Brazilian oilfields

ATLANTIC OCEAN

BRAZIL

Brasilia

Rio de Janeiro

São Paulo

8 bn barrels

Oil resources (barrels)

Saudi Arabia 260 bn

Brazil 14.4 bn

UK 4 bn

GREENLAND: A NEW OIL FRONTIER?

Case study 9

The US Geological Survey believes that there could be 90 billion barrels of oil and 50 trillion m³ of gas in the Arctic region, including off the southwest coast of Greenland, where Cairn Oil (a Scottish exploration company) may have located oil (see Figure 4.11). For the Greenland people, this represents a chance to move away from dependence on fishing, tourism and handouts from the Danish state as Denmark currently owns the island. The islanders see oil/gas as helping Greenland to become independent.

Opposition

Opposition to drilling has grown since the environmental disaster of *Deepwater Horizon*, particularly as extreme weather conditions could add to the risk of an oil spill in the

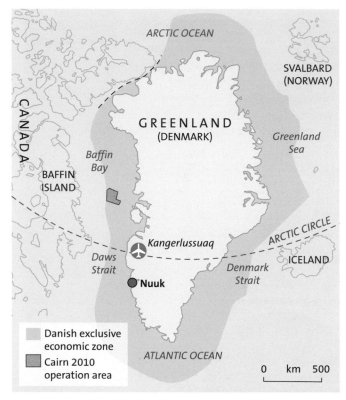

Figure 4.11
Greenland and the location of the Cairn Oil operation area

fragile Arctic environment, where oil would break down much more slowly than in the warm Gulf of Mexico. Another problem is the impact of climate change, which has led to widespread melting of the Arctic ice sheet. On the one hand it has improved shipping and port access but on the other hand floating icebergs are a constant danger and need to be guided away from the exploration rigs.

Greenpeace, the global environmental campaign group, has dispatched its ship *Esperanza* to highlight concerns. The Greenland authorities point out that the water depths are only around 450–500 metres, and that safety precautions are extremely rigorous, with 16 vessels working on standby around the Cairnwell rig to maintain safety standards.

The secretary general of Greenpeace Norway claims that drilling in Greenland's icy waters is very much a last resort and shows how the oil industry is being forced to the final frontiers by exploiting deep water oil in pristine, yet hostile, environments. However, the Greenlanders are keen to proceed with new exploration and are licensing areas for 2011 and 2012. They see the Greenpeace protest as a threat to their economic wellbeing and claim they 'cannot live by fish alone'. Note that the costs of producing oil in the Arctic are almost twice those in Middle Eastern areas.

Similar questions — 'wildlife sanctuary or oil bonanza?' — are being asked about many areas in the Arctic, including the Alaskan National Wildlife Refuge (ANWR) where there has been intense pressure to drill since the 1970s. The estimated 16 billion barrels of oil are tempting because they would improve the USA's energy security, but this could be only 6 months' supply for the USA at current usage rates. Whereas President Bush was very pro-drilling, President Obama has given permission for offshore drilling only.

6 Question

'The arguments for and against drilling for oil in high-quality environments are complex'. Discuss.

Guidance

Start with Greenland and then look at Alaska. Use www.anwr.org and www.defenders.org (type ANWR in the search box) to develop your arguments.

Issues include an assessment of the ecological and environmental value, as well as the economic value of the oil (how long would it last?) There is also the issue of the range of environmental damage.

REVIVING IRAQ'S OIL INDUSTRY

Increasing Iraq's oil production from the current 2.5 million barrels to a planned 12 million barrels a day, will probably take up to 8 years because of Iraq's continuing instability and ongoing political violence. The oil infrastructure, especially the pipelines, has become a focus for insurgent attacks in remote desert regions. Trade union leaders have called for strikes over oilfield wages, and tribal associations (largely Kurds) are asking for compensation of $1 a barrel as they claim they have lost their ancestral tribal lands to oil rigs. Both will push up costs.

There is also opposition from some politicians to foreign oil companies returning, but their expertise is seen by others as vital to sorting out the logistics. The most immediate problem is how to import vast quantities of equipment without swamping partly damaged ports such as Umm Qasr. Expansion will be difficult as the port's outlet is in disputed waters (Iran) and the port is a haven for Iraqi smugglers, who the new regime is struggling to control. Another adverse consequence of the war is that many fields such as Rumaila are infested with landmines.

At stake is Iraq's recovery, using the revenues from six large oilfields containing around 45 billion barrels. However, there are many obstacles to overcome before oil can start pumping again.

Unconventional oil sources

Figure 4.12 locates the main sources of unconventional oil. Superficially, there are huge supplies — more than enough to postpone peak oil for around 40 years, but at what cost? Unconventional sources herald a new era of expensive, technologically demanding and environmentally damaging oil production. *Case study 11* gives details of one such source — Canada's tar sands.

Figure 4.12
The main sources of unconventional oil

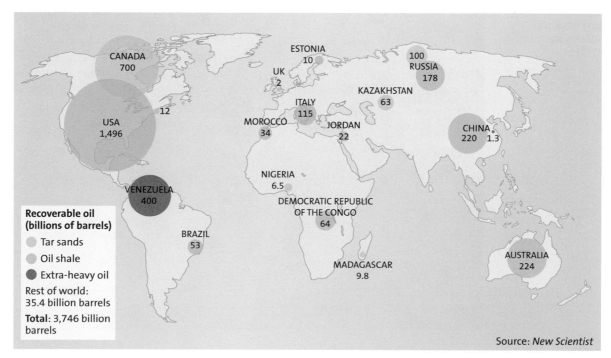

Source: *New Scientist*

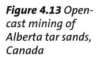

The most famous non-conventional resources are the Canadian tar sands, where proven reserves are second only in size to the giant conventional Saudi Arabian fields. Producing crude oil from the tar sands, found beneath more than 140 km^2 of prime forest in northern Alberta — an area greater than the size of England and Wales — is considered by many to be an environmental disaster for a number of reasons.

Tar sands containing bitumen are extracted from huge open-cast mines (to keep costs down), which scar the landscape. Millions of tonnes of ecosystems and topsoil are scooped away, and millions of litres of water are diverted from rivers (up to five barrels of water are needed to produce a single barrel of crude). At a current production rate of 1.2 million barrels per day, the industry is already reaching the legal limits as to what can be drawn from the Athabasca River in winter. The water used in the process becomes so contaminated that it has to be stored in ponds.

Another big problem is that production of oil from tar sands is extremely energy intensive. The booming oil-sands industry will produce 100 million tonnes of carbon dioxide equivalent, ensuring Canada will miss its Kyoto target. It takes about 29 kg of carbon dioxide to produce a barrel of conventional oil, but for tar sands this can be as much as 125 kg — so in greenhouse-gas terms this makes tar sands a very dirty fuel. It would also use substantial supplies of Canada's natural gas, which would have to be supplied to this remote area of Canada, involving considerable infrastructure development.

As well as these environmental concerns, there is the issue of costs. In a scenario most favourable to oil sands — oil prices at above $80 a barrel, a continued growth in demand, and a supportive regulatory environment — the analyst IHS CERA predicts that output from the Canadian tar sands could reach 6.3 million barrels per day by

Figure 4.13 Open-cast mining of Alberta tar sands, Canada

David Dodge, The Pembina Institute (www.pembina.org)

2035. However, that would require production to grow alarmingly in the face of substantial environmental opposition. Moreover, with the drop in oil prices in 2010 as a result of low demand in the recession, projects aiming to increase production to 1.7 million barrels a day were cancelled or put on hold.

Oil shales

Another source of unconventional supply is **oil shales** — sometimes described as 'oil from a stone'. Around 2.5 trillion barrels of 'oil equivalent' has been identified, with the largest deposits of all in the mountain west of the USA (Green River in Wyoming, Colorado and Utah). The rock is heated up to 500°C until the kerogen decomposes into a synthetic crude oil and a solid residue — traditionally by mining the shale and heating it in a huge oven. This is an expensive, energy-intensive process and leaves a huge volume of waste stored in unsightly slag heaps, known as bings in West Lothian, Scotland, which has produced oil from shale in the past.

What is needed is an in-situ production method that is less visually damaging to the environment, for example using microwaves or high-temperature gas injection to create an oil reservoir underground. This can then be extracted using conventional drilling. The IEC estimates that shale oil will cost $80–$100 per barrel to produce, with a possibility of a carbon emission tax to be added to this.

Shell has experimented with shale oil extraction at its development site in Cathedral Bluffs, Colorado, using in-situ technology. The big issue is that the process is power hungry, and therefore makes the shale a dirty fuel in terms of carbon dioxide emissions. Most experts expect no significant (i.e. commercial) shale oil production before 2030 — but both tar sands and shale oil are nice insurance policies for the future.

Conversion processes

There are two possibilities here: coal to liquids (CTL) and gas to liquids (GTL), but the estimated direct costs are high, ranging from $40 to $110 per barrel (see Figure 4.8).

Making liquid fuels from coal is nothing new — the technology was first developed by Germany during the Second World War, by a coal-rich, oil-starved country desperate for diesel and paraffin, and then later by South Africa during the sanctions imposed as a result of an Apartheid regime. South Africa has plentiful cheap coal and using the SASOL process has successfully run a CTL plant since the late 1980s, producing high-quality fuel.

The question is, can we really afford to use coal in this process? It takes 2 tonnes of coal and up to 15 barrels of water to produce a single barrel of synthetic oil. Moreover, because of feedstock and energy demands of the production process, CTL fuels have roughly double the carbon emissions of conventional crude, even if carbon capture and storage was introduced.

Liquid fuels can also be made from gas, of which there is currently a worldwide glut. GTL emits much less carbon than CTL, because the original feedstock is cleaner, but still more than conventional crude. Nearly 50% of the 280 m^3 of gas used to produce a barrel of GTL are burnt during the conversion process. Two large plants have recently been opened in Qatar and Nigeria.

In conclusion, will unconventional oil resources be able to fill the gap left by diminishing reserves of conventional crude oil? Certainly they will produce much dirtier oil and at higher costs. Globally non-conventional sources would be penalised by any carbon-pricing regime, perhaps adding $5 a barrel to tar sand fuel, $12.50 for GTL and $30 for CTL.

In a recent UK Energy Research Centre report it was stated:

> If everything goes well, oil sands might produce 6 million barrels per day in 20 years time, but by then we will need to add at least 10 times that much capacity — without allowing for any growth in demand. It is very hard to see non-conventional sources riding to the rescue.

Oil: benefit or curse?

High oil prices have made countries such as United Arab Emirates so wealthy that they are investing in shipping lines and port infrastructure, manufacturing companies, and even a football club (Manchester City). However, while the governments get wealthy from oil and gas revenues, frequently they fail to trickle down this wealth to improve the health and welfare of their people.

One reason for this is that Western governments and TNCs have turned a blind eye to corrupt governments in the pursuit of energy security. The African continent is a source of many new discoveries of mineral and energy reserves. Conditions in many African states can best be described as chaotic, with countries such as Angola or Equatorial Guinea rated to be the most difficult working environments in the world. In a continent noted for widespread corruption, the challenge for governments in these countries is to raise the capital and commandeer the technology to develop their oilfields.

In Ghana, the government has vowed to be a model oil investor, using oil revenues in modernising agriculture, investing in technology and infrastructure, and diversifying the Ghanaian economy. Others, including Angola, have established a relationship with China, using Chinese efficiency and expertise to develop the oil, which will ultimately be sold back to China. Uganda has found oil on the shores of Lake Albert, but will need help with pipelines and infrastructure to extract the oil for export.

The case study of the Niger Delta region shows how 'bad' development can make oil as much a curse as a blessing.

Case study **12**   **THE CURSE OF NIGERIA'S OIL**

With 606 individual oil wells, the Niger Delta provides 80% of the Nigerian government's revenue and 95% of Nigeria's export earnings. The field is exploited primarily by the TNCs Shell and Exxon Mobile in partnership with the Nigerian government. The TNCs provide infrastructural development and expertise to exploit the oil. The relationship has been a turbulent one, with the TNCs being expelled for several years. Nigeria's vast oil and gas reserves are currently the biggest in Africa — so the question is, how could all this potential oil wealth be a curse?

Oil spills

Nigeria was in African terms very much an oil pioneer. Developments began in the late 1950s and early 1960s, so many of the pipelines and pumping stations are now ageing. Pollution is rife and over 300 oil spills have occurred in the last 5 years. A 3-year UNEP investigation costing £6.5 million, admittedly paid for by Shell, exonerated the company from blame for 90% of the spills, i.e. only 10% of oil pollution was caused by equipment failure and oil-company negligence.

The main cause of the widespread oil spills, the report concluded, was local people illegally stealing oil and sabotaging company pipelines. In 2009 Amnesty International calculated that an equivalent of 9 million barrels was spilt — far more than the *Deepwater Horizon* accident.

Political problems

Ever since the opening up of the oilfields, when Biafra, which was made up of oil-rich provinces, tried to break away from Nigeria, there has been political turmoil in the delta zone, Ogoniland. In 1995 Ken-Saro-Wiwa, a leading poet and ecological activist, and eight other Ogoni leaders were hanged by the Nigerian government after leading a peaceful uprising against pollution caused by Shell and Exxon — leading to a suspension in production.

Since 2005, the delta creeks where the bulk of Nigeria's vast oil and gas reserves are found have been at the heart of a violent campaign by local tribespeople against the Western TNCs and the Nigerian government. Calling for a bigger share of the vast revenues for local people, from the highly priced high-grade crude oil, gangs attacked foreign oil groups' pipelines and kidnapped hundreds of expatriate staff.

In 2009 the then president negotiated an amnesty with the militant groups that carried out the attacks — especially MEND (Movement for Emancipation of the Niger Delta). The ambushes had resulted in a drop in Nigeria's oil production and caused oil revenues to fall by 25%. The only way the delta could function was by employing thousands of private security guards. The president developed a scheme that persuaded thousands of young local people to swap weapons for cash, giving them monthly payments of $430 and training them for employment. The new president, Goodluck Jonathan, who comes from the delta zone, invested in a scheme under which rebels could have counselling and discuss job prospects, but unfortunately the project has foundered and the disputes have reignited.

Pollution

The lives of the Ogoni people who live in the delta zone are threatened by pollution from oil. These victims include fishermen and subsistence farmers, who have pipelines crossing their land. Land and water have been degraded by oil and air quality lowered by the flaring off of gas from the oil wells. Life expectancy in the areas is reduced by the pollution (around 40 years in the region) and many communities lack a clean water supply.

The June 2009 report by Amnesty International called the extensive damage caused by oil spills in the delta 'a human rights tragedy', especially as local people get so few benefits from the massive oil revenues.

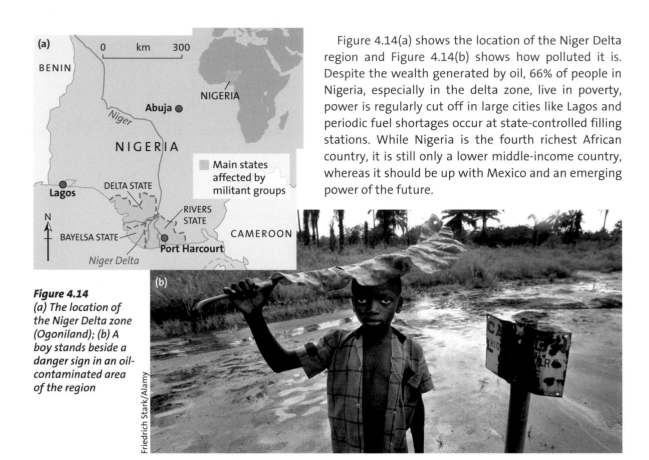

Figure 4.14(a) shows the location of the Niger Delta region and Figure 4.14(b) shows how polluted it is. Despite the wealth generated by oil, 66% of people in Nigeria, especially in the delta zone, live in poverty, power is regularly cut off in large cities like Lagos and periodic fuel shortages occur at state-controlled filling stations. While Nigeria is the fourth richest African country, it is still only a lower middle-income country, whereas it should be up with Mexico and an emerging power of the future.

Figure 4.14
(a) The location of the Niger Delta zone (Ogoniland); (b) A boy stands beside a danger sign in an oil-contaminated area of the region

Friedrich Stark/Alamy

Using case studies

Questions

1 Explain why the case study is entitled 'The curse of Nigeria's oil'.
2 What steps could be taken in Nigeria to make oil become less of a curse?

Guidance

1 Mention all environmental impacts, the abuses of human rights and the inequalities within Nigeria. The following sites reflect the high level of controversy surrounding the Niger Delta:
   http://en.wikipedia.org/wiki/Conflict_in_the_Niger_Delta
   www.globalissues.org/article/86/nigeria-and-oil
   www.cfr.org/publication/22809
   http://ngm.nationalgeographic.com/2007/02/nigerian-oil
2 Investment in the region to overcome poverty. Ensure TNCs employ a designated percentage of locals. Develop strict environmental safeguards. Update ageing oil infrastructure.

So far in this assessment of a world trying to avoid running out of oil the emphasis has been on developed nations seeking to achieve energy security. Developing nations are also dependent on oil, possibly more so as they may have few alternatives, and their industries and transport systems are less efficient. The developing nations pay a high price for oil, even today. They often lack bargaining clout, and because they need smaller quantities they cannot take advantage of economies of scale, especially as their ports can only accommodate small tankers.

The nuclear question

People have been discussing the pros and cons of nuclear power for a long time. In the past few years nuclear power has once again surfaced as a 'hot topic' on the global political agenda. There are a number of reasons for this:

- global energy insecurities
- growing energy demands and changing standards of living, especially in the middle-income emerging economies
- the need to meet the EU's post-Kyoto targets for emissions reduction
- states such as Iran using nuclear power as a way of flexing their geopolitical muscles
- concerns over peak oil and the price of fossil fuels (see page 54)

It is clear that in many parts of the world attitudes towards nuclear power are shifting in response to climate change and fears over the longevity of traditional fossil fuel supplies. Figure 5.1 shows global nuclear energy production and its relative share of total electricity production. There has been a substantial increase in nuclear power output worldwide, but since the early 1990s its share of total electricity production has declined. In 2007 nuclear power accounted for 6.3% of the world's total primary energy supply.

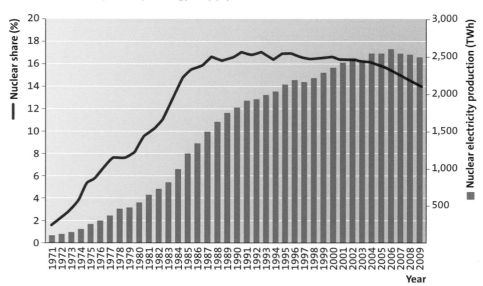

Figure 5.1
Nuclear energy production and share of total electricity production

Figure 5.2 shows how nuclear generation varies on a global scale. The most up-to-date figures available from 2007 show USA as the largest producer with 816 TWh of capacity (29% of the world total). However this only supplies 19% of the country's demand for domestic electricity. France on the other hand has almost 80% of its domestic electricity supplied by nuclear power, but only contributes 16% of the global total.

Figure 5.2
Nuclear electricity generation by country, 2007

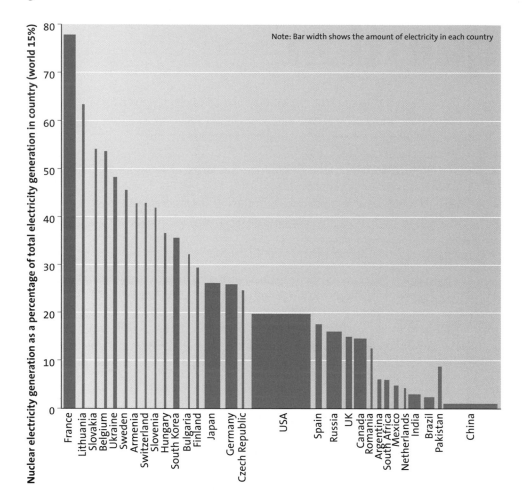

Folklore and fantasies?

One of the key problems with the nuclear question is trying to uncover fact from fiction. There are many controversial sources and arguments. Table 5.1 gives some of the most important anti-nuclear criticisms and some pro-nuclear responses. To what extent are the criticisms and the responses valid?

Table 5.1 *Anti-nuclear criticisms and pro-nuclear responses*

Anti-nuclear argument	Pro-nuclear response
Uranium mines pollute the environment, e.g. tailings dams cause pollution through leakage.	Uranium mines today aim for zero emission of pollutants. Any water release is of surface runoff and is close to drinking standard. Tailings retention does not normally cause pollution offsite.
Nuclear wastes (i.e. spent fuel) are an ongoing and unresolved problem.	In all countries using nuclear energy there are well-established procedures for storing, managing and transporting such wastes, funded by electricity users. Wastes are contained and managed, not released. Storage is safe and secure, and plans are well in hand for eventual disposal.
Nuclear reactors cannot be guaranteed safe. Chernobyl was an inevitable accident and resulted in a large death toll.	The nuclear industry has an excellent safety record. Some Soviet-designed and built reactors have been a safety concern, but are much better now than in 1986. The Chernobyl disaster was irrelevant to any modern reactor that could be built today. According to UN figures, the Chernobyl death toll is 56, including those who died from cancer.
Nuclear reactors are vulnerable to terrorist attacks like 9/11. Waste and spent fuel storage is even more so.	New reactors are strong, with robust pressure vessels and internal structures. Evaluations since 2001 suggest that power reactors would be well equipped to survive a 9/11 type impact without any significant radioactive hazard locally. Civil waste and spent fuel storage is also robust and often below ground level.
Decommissioning nuclear plants will be too expensive to undertake.	Decommissioning is usually funded while the plant is operating. Experience to date gives a good idea of costs and earlier estimates are being revised downwards.
We cannot be sure that our uranium does not end up in weapons, e.g. in France or China.	Safeguards would detect any diversion. Today military materials are being released for dilution and civil use, so there is not even a reason for diversion in customer countries such as France and USA. France no longer has the means to enrich uranium beyond reactor grade. China has ample uranium for any military programme, but is understood to have ceased putting uranium into this in the 1990s.
Nuclear energy makes only a small contribution to world energy needs.	Electricity generation uses 40% of the world's primary energy. Nuclear provides around 16% of world electricity — more than the total electricity produced worldwide in 1960.

Source: World Nuclear Association

How green is nuclear power?

Worries about emissions from the burning of fossil fuels to generate electricity have led many leading environmentalists to lobby for nuclear power, including Stephen Tindale, a former director of Greenpeace. Now pressure is mounting on governments and individuals to find low-carbon solutions to generating electricity. Supporters of nuclear power claim that in generating electricity it produces very little carbon dioxide. This statement, however, is somewhat simplistic and should be taken against the backdrop of the real life cycle carbon dioxide emissions of electricity generation (see Figure 3.2 on page 40).

As you will have found in Part 3, calculating carbon footprints is problematic and contentious. Some studies have shown that nuclear power generation has a relatively small carbon footprint (around $5\,g\,CO_2$eq per kWh). Other studies in 2006 and 2007 respectively, showed that the nuclear footprints were much larger, at 84 and 93 g CO_2eq per kWh. Earlier research has been criticised for not looking at the whole carbon footprint in a true life-cycle analysis. This makes it difficult to confirm the real footprint of nuclear power against other sources of electricity.

For some parts of the world, looking towards a nuclear future has wide ranging benefits; in 2010 over 30 countries were actively considering embarking upon nuclear power programmes. These ranged from sophisticated economies to middle-income and developing nations (Figure 5.3).

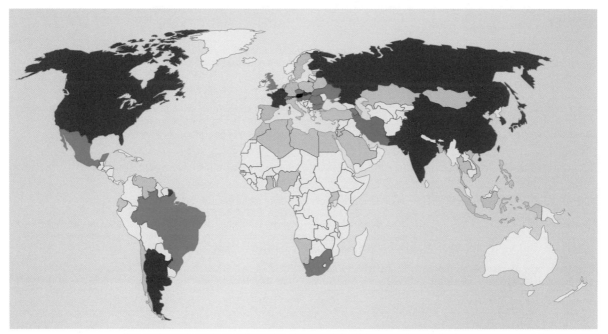

Figure 5.3
The status of nuclear power globally in 2009

■ Operating reactors, building new reactors
■ Operating reactors, planning new build
■ No reactors, building new reactors
□ No reactors, planning new build
▨ Operating reactors, stable
■ Operating reactors, considering phase-out
■ Civil nuclear power is illegal
□ No reactors

Italy

In Italy, for example, electricity consumption in 2005 was 330 billion kWh, giving per capita consumption of 5,640 kWh per year. Then, the country's electricity energy mix was 50% gas, 15% oil, 16% coal and 14% hydroelectric. Imports accounted for 50.3 billion kWh (about 15%), mostly via French nuclear power. Due to the high reliance on oil and gas, as well as imports, Italy's electricity prices are 45% above EU average. Italy is now the only G8 country without its own nuclear power, and is the world's largest net importer of electricity. In May 2008 the Italian government confirmed that it would build a number of nuclear power plants within 5 years. It will work towards having 25% of its electricity from nuclear power by 2030, which will require the construction of eight to ten large reactors.

China

Other countries have different motives to develop their nuclear electricity capacity. In 2010 mainland China had 12 nuclear power reactors in operation, 24 under construction, and more about to start construction. Additional reactors are planned, including some of the world's most advanced, to give more than a tenfold increase in nuclear capacity to 80 GW (gigawatts) by 2020, 200 GW by 2030, and 400 GW by 2050.

China is rapidly becoming self-sufficient in reactor design and construction, as well as other aspects of the nuclear fuel cycle. This is a key benefit as knowledge of reactor design and production is highly specialised and valuable, especially as the country becomes less reliant on overseas input. Such expertise could form an important knowledge export as part of China's network of trade relations with Africa and Latin America.

Nigeria

In 2008 Nigeria's per capita consumption of electricity was only 113 kWh per year but it is growing. Nigeria is Africa's most populous country (about 140 million) but suffers shortages of electricity. Inadequate power supplies and power shortages have caused some industries to relocate to Ghana. This is in spite of Nigeria being a leading producer of both oil and gas (see Part 4). Recently Nigeria has embarked on a new nuclear programme and will commence electricity generation through nuclear energy by 2019 with an initial capacity of 1,000 MW. Nigeria has been partnering the International Atomic Energy Agency and other African countries in this quest; this has boosted knowledge sharing in nuclear science and technology.

8 | Using case studies

Questions
1. Explain why the three countries in the case study have chosen to develop nuclear power.
2. What are the downsides of nuclear power development?

Guidance
1. Think about energy security as well as impending electricity gaps and scientific prestige.
2. General issues of safety, radioactive waste disposal, costs etc. and military concerns in unstable countries, for example Pakistan in 2010

The economics of nuclear power

From nuclear power's commercial beginnings, the promise of cheap power (infamously, 'power too cheap to meter') has been one of the main claims of the nuclear industry. Supporters of nuclear power today still insist that it is cost-competitive with other forms of electricity generation, except where there is direct access to low-cost fossil fuels, such as coal. The trouble is that, while the small amounts of fuel are relatively cheap, nuclear power plants are expensive to build and the pay-back from the initial investment is very slow. It is also hard to know the true completed cost of a modern nuclear plant since most of the operational power stations were built a long time ago when the economics were different.

Figure 5.4 shows the complexities of nuclear power costs, based on fixed and operational costs.

Figure 5.4
(a) Fixed or capital costs: factors that determine cost per kilowatt per hour (b) Variable or operating costs: costs that vary according to the amount of electricity produced

(a)

Construction cost and time
Past costs are the best indicators of future costs, but these data are often not available in a proper audited format. Large projects often experience unforeseen costs.

Cost of capital (borrowing)
This varies from country to country and is strongly influenced by inflation and the credit-rating of both the company and the host country.

Reliability of the plant
A good measure of reliability is how effective it is at producing electricity for sale (when taking into account breakdowns and other interruptions). This is an important part of the fixed cost. Operational efficiency varies worldwide from about 60%–85%.

Decommissioning costs
Difficult to calculate and often initially underestimated. Many government subsidies have been withdrawn to reduce taxpayers liability.

Insurance and liability
Complex arrangement exists for liability and insurance, but at an average they account for about 10% of generation costs.

(b)

Non-fuel operations and maintenance cost
Often not used in accounting, they related to increased staffing costs to achieve greater reliability. With increased emphasis on safety and inspection these can be twice fuel costs.

Accounting lifetime
Increasing the life span does reduce fixed costs. Many new plants should operate for 60 years. But there are unknowns with replacement of worn-out equipment and other problems down the line.

Fuel cost
Historically very low the cost of uranium is now on the increase, but many reactors use reprocessed fuel which is still comparatively cheap.

Liberalisation of energy markets
Nowadays building nuclear power stations tends to be very risky with no government support (tax payers used to pick-up the tab if things went wrong). Companies such as EDF bear all the risks.

The most up-to-date figures suggest that costs are exceeding original estimates. A reactor in Finland, for example, originally priced at €3 billion is 3 years late and is expected to cost nearer €5 billion. A similar picture is found in France with a new reactor costing more than 20% over the original estimate. Equivalent coal and gas-fired power stations cost around €1 billion.

Another worry is the possible increase in cost of uranium over the longer term (see Figure 5.5). Canada and Australia between them have nearly 50% of the world's uranium — most is extracted by a handful of companies (state and private) in very few places. This also increases the supply risk. The nuclear fuel cycle depends on complex processing technology and there is a mismatch between the location of uranium deposits likely to be mined in the future and the location of facilities needed to process the raw material. The hazardous nature of uranium and its movement within this fuel cycle is logistically difficult and politically sensitive.

Nevertheless, both the fixed (capital) and operating costs of nuclear power look competitive when viewed against other types of energy generation (see Figure 5.6). But it is extremely difficult to provide a balanced assessment of the true cost of nuclear power as different studies tend to contradict each other.

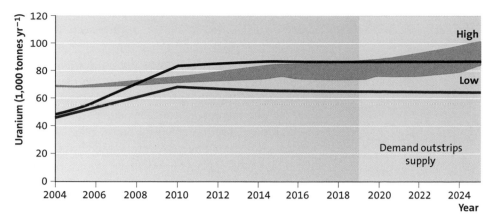

Figure 5.5
Uranium shortage

What is certain is that in liberalised electricity markets (see page 101), the high initial up-front costs of building a new reactor plant make all but the largest privatised electricity companies reluctant to invest in the infrastructure. In many countries, therefore, the responsibility for nuclear development rests with the state. Nuclear energy's best hope may come in the form of carbon pricing, which forces fossil-fuel plants to pay for the environmental cost of the carbon dioxide they release. The price under Europe's emissions-trading scheme is about £12 per tonne (2010 prices). This is far short of what the nuclear industry would like, at nearer £40 per tonne, to make its plants look more attractive.

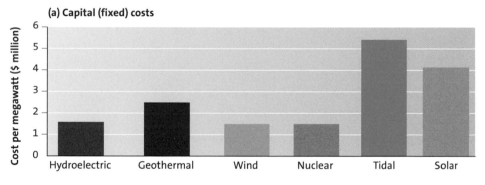

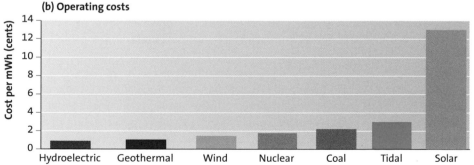

Figure 5.6
Relative operational and fixed costs of nuclear against other energy types

Figure 5.7
The UK's nominated sites for new nuclear power stations

Legend:
- Current sites
- Shut-down sites
- New sites

Map labels:
SCOTLAND
Hunterston
Torness
Chapelcross
Braystones
Calder Hall
Sellafield
Hartlepool
Kirksanton
Heysham
Wylfa
ENGLAND
Trawsfynydd
WALES
Sizewell
Bradwell
Berkley
Oldbury
Hinkley Point
Dungeness

A White Paper in 2008 made it clear that the UK government supported new nuclear power stations and that they should have a role to play in this country's future energy mix alongside other low-carbon sources. Nuclear power was described as a 'proven source of low-carbon energy'. This drew criticisms from environmentalists who warned of a problematic legacy of radioactive waste. However, in November 2009 the government approved ten sites in England and Wales (see Figure 5.7). Many of the proposed sites are already home to nuclear power plants. It is hoped that some of the locations will be operational by 2018.

Such new developments are always controversial and become political headaches, especially with all the associated issues of nimbyism. However, an independent survey in 2008 by consultants Accenture revealed that people in the UK were generally positive about nuclear power. In particular their findings showed the following:

- 88% of respondents felt it was important that the UK reduced its reliance on fossil-fuel-generated power.
- More than half the respondents felt that the UK should increase its nuclear power generating capacity.
- In the period 2003–08, support for nuclear power had increased by 30%.

A more in-depth analysis of the results reveals other interesting patterns. When asked about increasing nuclear power-generating capacity (Figure 5.8) there was a big difference in responses from males and females, although no significant difference between regions or socioeconomic categories. The reasons for the biggest resistance to nuclear were centred on issues of decommissioning and waste handling. People were also worried about the operational safety of nuclear plants and the high costs of building them.

Figure 5.8
Nuclear attitudes survey: 'Do you think that the UK should increase its nuclear power-generating capacity?'

		General population	
		Yes	No
Gender	Male	68%	32%
	Female	38%	62%
Age	18–34	54%	46%
	35–55	44%	56%
	55+	60%	40%

The UK coalition government elected in 2010 gave assurances that it would support the Labour nuclear programme providing there was no public subsidy involved. This means that most of the ten new plants are likely to go ahead providing they can be privately financed and the planning applications are approved. The first ones will not be ready until at least 2018, but will make a major contribution to base-load electricity. The building on existing nuclear sites is likely to diminish local opposition as the new nuclear power stations will secure jobs that would probably have disappeared as a result of decommissioning.

Question

A survey represents a snapshot in time. What factors could lead to a change in public support for a UK nuclear renaissance?

Guidance

Think about comparative costs of other fuels (gas glut), increasing concerns over risks from terrorism and explosions, or diminishing support for doing something about climate change and its impacts.

ASIA'S NUCLEAR ENERGY GROWTH

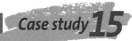

Case study **15**

In contrast with North America and most of western Europe where growth in electricity-generating capacity and particularly nuclear power levelled out for many years, a number of countries in east and south Asia are planning and building new nuclear power reactors to meet their increasing demands for electricity. The World Nuclear Association states that projected new generating capacity in this region involves the addition of some 38 GW per year up to 2010, and 56 GW per year from 2010 to 2020, with up to one third of this replacing retired ageing coal and nuclear plants. This is about 36% of the world's new capacity (current world capacity is about 3,700 GW, of which 370 GW is nuclear). Much of this growth will be in China, Japan, India and Korea, which are all technologically advanced countries.

Rapidly rising populations and industrial growth combined with major concerns about energy security have put energy at the top of government agendas in the region. Power demand is growing more rapidly in east and southeast Asia, from Japan around to India, than in the Western world.

India

In India for instance, the technology for the manufacture of various components and equipment for new reactors is now well established and has evolved through active collaboration with other hi-tech Indian industries. India has a flourishing and largely home-grown nuclear power programme and expects to have 20,000 MW of nuclear capacity on line by 2020 and 63,000 MW by 2032. India is planning to increase the contribution of nuclear power to overall electricity generation capacity from 4.2% to 9% within 25 years.

India is outside the Nuclear Non-Proliferation Treaty due to its weapons programme. This means that it has been excluded from trade in nuclear plant or materials. Because of these trade bans and the lack of indigenous uranium, India has been developing a nuclear fuel cycle to exploit its reserves of thorium. It is working towards becoming a world leader in nuclear technology due to its expertise in fast reactors and the thorium fuel cycle.

Japan

Despite having suffered the devastating effects of nuclear weapons in wartime, Japan has embraced the peaceful use of nuclear technology to provide a substantial portion of its electricity. Figure 5.9 illustrates the transition in the share of power sources for Japan. Note that the big increase in nuclear power has been paralleled by the increased use of coal as a temporary solution to the 'energy gap'.

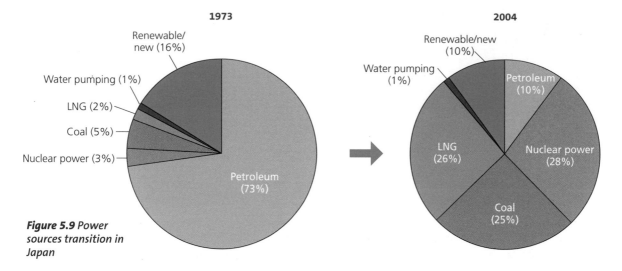

Figure 5.9 *Power sources transition in Japan*

As Japan has few natural resources of its own, it depends on imports for some 80% of its primary energy needs. Initially it was dependent on fossil fuel imports, particularly oil from the Middle East (oil fuelled over 70% of the electricity in 1973 — Figure 5.9). This geographical and commodity vulnerability became critical following the oil shock in 1973 and a review of the domestic energy policy resulted in diversification, including a major nuclear construction programme. A high priority was given to reducing the country's dependence on oil imports. Today, nuclear energy accounts for almost 30% of the country's total electricity production (2010). There are plans to increase this to 41% by 2017, and 50% by 2030. This will help with both energy security and emissions reduction.

10 **Using case studies**

Question

Produce a third pie chart for 2025 for power generation in Japan. Annotate it to justify your predicted energy mix.

Guidance

You can use the case study to put in a likely nuclear percentage. You can then look at the balance between fossil fuels and renewables. Remember that Japan is highly energy insecure so using renewables such as hydropower and wind may be a good solution. Equally, post-Kyoto targets may lead to a rapid drive towards cleaner fuels.

Nuclear energy is currently experiencing a renaissance for a whole range of reasons. It would only take another Chernobyl to halt this surge. On the other hand, technological developments such as nuclear fusion may make nuclear power much more cost-effective.

Renewable opportunities

Renewable energy is one of the fastest-growing sectors in the world economy. Renewables can generate economic growth and provide local jobs. They have the potential to contribute towards electricity generation, hot water and space heating, and transport fuels. Can renewables really provide an answer to our future energy needs?

Renewables: a universal panacea?

On paper, the amount of recoverable renewable energy available globally would serve our needs several times over. So could we meet all our energy needs from renewable sources? The answer is complex, and there are many challenges.

Challenge 1: Intermittency

The more electric energy supplied by renewables, the more unstable national electricity grids become. Renewable generators only produce electricity intermittently. The output of a solar photovoltaic plant can drop from 50 MW to close to zero in seconds if a cloud passes over the sun, and then surge up again when the sky is clear. As more renewables come online it becomes more complex to manage fluctuations in the grid. This needs to done by the end users so that smart grids, along with stores of electricity, become a part of the solution.

Challenge 2: Cost of renewables vs conventional sources

We have already seen in Part 5 the relative cost of nuclear power against other forms of electricity generation. Look again at the graphs on page 75. At present many renewables are considered too costly when compared with conventional fossil fuels, especially gas and coal used for electricity. There may also be huge outlays for large renewables projects (e.g. the Severn Barrage), which are inevitably controversial.

Challenge 3: Future changes

A really big challenge is that we do not know what the future holds in terms of demands for electricity (both industrial and domestic) or in terms of transportation (for instance it seems unlikely that aircraft will be moved away from fossil fuels in the foreseeable future). How will changing energy attitudes, prices, efficiency and technology alter spatial and temporal patterns of demand?

Challenge 4: The biofuels issue

This is another complex debate, especially when set against issues linked to the burning of fossil fuels and targets to reduce emissions. This topic is explored in detail on pages 83–85.

Challenge 5: Local opposition, habitat destruction and space inefficiency

There may be issues with the amount of space required to either produce or store the renewable energy. For example, this is particularly problematic with biofuels, where they are displacing food crops. Wind farms are often criticised for blighting the landscape (although this is very much a matter of opinion).

Challenge 6: Limited availability

Some resources, such as solar energy and wind energy, are only abundant in certain parts of the globe at limited times. Technically it is possible to export electricity through a worldwide smart grid system but there are many issues with this, including cost of infrastructure and the end user paying for energy. There are few hydrogen cars on the market because of the transport and storage costs of the cells.

Despite all the challenges presented by a shift towards renewables, there are a number of countries that can be classified as renewables 'champions' (Figure 6.1).

Figure 6.1
Renewables 'champions', 2009

Existing capacity at end of 2009	Global rank				
	1	**2**	**3**	**4**	**5**
Renewables power capacity (including only small hydro)	China	USA	Germany	Spain	India
Renewables power capacity (including all hydro)	China	USA	Canada	Brazil	Japan
Wind power	USA	China	Germany	Spain	India
Biomass power	USA	Brazil	Germany	China	Sweden
Geothermal power	USA	Philippines	Indonesia	Mexico	Italy
Solar PV (grid-connected)	Germany	Spain	Japan	USA	Italy
Solar hot water/heat	China	Turkey	Germany	Japan	Greece

The pros and cons of wind

It is likely that an ever-increasing number of wind farms will become part of the energy mix, both at home and overseas. But are wind farms really a good energy solution for the future? Table 6.1 shows some of the arguments presented for and against.

For...	Against...
'They really are needed.' Increases in the demand for electrical energy together with issues of energy insecurity and worries about global warming mean that wind power is a good option. It can be installed at a range of scales from individual and community, to large-scale, offshore plants.	'They are damaging to landscapes.' This applies both to local people and to tourists who may visit the area. Opponents talk about destruction of the landscape character and that in many countries planning legislation is either being overlooked or cannot cope with the rush of wind farm applications.
'They bring benefits to the local economy.' The wind industry creates many new jobs: globally, over 400,000 people are now employed in this industry, and that number is expected to be in the millions in the near future. Wind generates income in the form of tax revenues, lease payments and capital investment. There is also revenue generated from planning, development, and construction.	'They are harmful to birds and bats.' This is highly controversial and very much depends on size and location of the turbines. Some research suggests large numbers of fatalities: in Wolfe Island, Canada, between July and December 2009, 602 birds and 1,270 bats were reportedly killed by the turbines. This is likely to be unusually high, rather than the norm.
'They reduce carbon dioxide emissions.' When operational, most renewables have low or non-existent carbon dioxide emissions. However, the development of wind farms does have carbon 'costs' in the manufacture of the turbines themselves, plus costs of installation (including transport) and operation.	'Noise can be a worrying issue.' Possible health effects from the low-frequency noise, e.g. headaches and earaches. However, these claims can be over-exaggerated. Personal damage/injury is difficult to measure.
'They are becoming cheaper to install and operate.' (Some) recent estimates put the cost per kilowatt hour at 4–8 pence, which is comparable with electricity from gas and nuclear power.	'They devalue property nearby.' Research in the USA and Britain suggests that there is no consistent relationship between wind turbines and property prices. But if enough people expect a negative effect, the fear will be self-fulfilling.
'They are accepted by most people.' Wind turbines represent a visible form of renewable power. Recent UK research found that they are favoured by four out of five people.	'They are very inefficient.' In the UK (the windiest country in Europe) some data suggest that a wind turbine produces less than 25% of its installed capacity (its power rating). The power it does generate is intermittent and unreliable. Downstream wind turbines may lose 20 or even 30% of their power compared with those at the front.

Table 6.1 Arguments for and against wind power

Overall the signs are good for wind energy. The Global Wind Energy Council predicts that the global wind market will grow by over 150% between 2010 and 2012, to reach 240 GW of installed capacity. This will account for about 3% of global electricity production in 2012 — a trebling from 1% in 2007.

In 2009, the USA was the world's largest producer of wind energy, generating over 35,000 MW and employing over 85,000 people (see Figure 6.2). China, which has succeeded in doubling its installed wind capacity every year since 2004, is not far behind. The Chinese Renewable Energy Industry Association forecasts that it will reach around 50,000 MW by 2015. The big challenge for China is internal transmission of electricity due to the huge geographical separation between the wind farms of the interior and major centres of power and consumption in China's coastal industrial heartland.

Figure 6.2
Installed wind capacity as a percentage of global total (end of 2009)

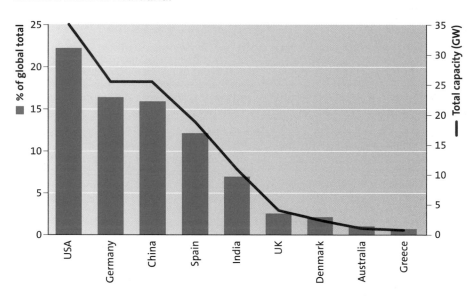

In most cases wind energy will divide communities, particularly at the local scale. But wind does not always lead to conflict or political problems. In several countries where it flourishes — such as Germany, China and Spain — the technology is relatively free from controversy, either because the public have been convinced of its worth, or people are simply prepared to accept top-down decisions.

Big dams: are they big trouble?

People have diverted water from its natural channels throughout history. Remains of water storage dams are found in Jordan, Egypt and other parts of the Middle East dating back at least 5,000 years. The era of large dams or 'megadams' started in the early part of the last century and has helped communities and economies harness water resources for food production, energy generation, flood control and domestic use. Current estimates suggest that some 30–40% of irrigated land worldwide now relies on dams and that hydroelectric power from dams generates around 16% of world electricity. Twenty-four countries depend on dams for 90% of their electrical power supply. But in recent years, the building of any dam has rarely escaped environmental, social or political controversy.

Given its long history and large scale, hydroelectric power (HEP) is one of the most mature of the renewables industries. In developed markets such as the EU,

USA, Canada, and Japan, where many HEP facilities were built 30–40 years ago, the industry is focused on re-licensing and re-powering as well as adding hydro-generation to existing dams. In emerging nations such as China, Brazil, Ethiopia, India, Malaysia, Turkey and Vietnam, utilities and developers are focused on new hydroelectric construction.

In recent years, China has had the greatest growth in hydropower capacity, nearly doubling its output during the 5-year period of 2004–09. It now has approximately 200 GW of installed capacity.

Those opposed to large dams can gather together a range of criticisms based on those already built, which have provided some benefits but have without exception destroyed river environments and uprooted the human communities that depended on them. There is also evidence that, even in the best examples, large dams tend to encourage population growth and resource extraction in areas that are marginal for human settlement and thus in the long run only worsen existing environmental problems and create new ones. Yet large dams in many parts of the world are part of the World Bank's development strategy. For example, the World Bank has provided more than $1 billion in loans for the Ertan dam in Sichuan, China's second-largest hydro project. China needs to provide power for its growing population and growing economy. Development is strongly linked to a reliable supply of energy.

Demystifying the biofuels and biomass debate

Biofuels have been put forward by some as the next big green investment opportunity. They will almost certainly be part of our low-carbon energy mix but are they really the 'green fuel' of the future?

Biofuels are produced from biomass — either living matter or waste from living matter. This includes crops, such as sugar cane and corn, that are grown specifically for fuel, as well as wood/chippings and animal or farm wastes. The term biofuels usually refers to liquid fuels, either bioethanol (ethanol) or biodiesel. Bioethanol is a form of alcohol often mixed with petrol, whereas biodiesel is a fuel in its own right (similar to oil-derived diesel) but is normally mixed with other fuels to make it more compatible with modern diesel engines. Figure 6.3 evaluates some of the different types of biofuel, both ethanol and biodiesel based.

The most recent data from 2008 reveal that within the EU Germany was the top consumer of biofuels (ethanol and biodiesel) with a total transport-based consumption of nearly 4 million toe. France was second with around 1.5 million toe and UK was much lower down the list at 0.4 million toe. For these countries, transport accounts for 20–25% of total carbon emissions. The EU Biofuels Directive suggests a 2020 goal of 10% transport use for biofuels. In Brazil, 16% of the country's total energy needs are supplied from sugarcane planted on only 1% of the country's arable land.

Crop	Used to produce	Greenhouse gas emissions (kg CO_2/mJ energy)	Use of resources during growing, harvesting and refining of fuel				Pros and cons
			Water	Fertiliser	Pesticide	Energy	
Corn	Ethanol	81–85	High	High	High	High	Technology ready and relatively cheap, reduces food supply
Sugar cane	Ethanol	4–12	High	High	Medium	Medium	Technology ready, limited as to where will grow
Switch grass	Ethanol	–24	Medium–low	Low	Low	Low	Won't compete with food crops, technology not ready
Wood residue	Ethanol, biodiesel	N/A	Medium	Low	Low	Low	Uses timber waste and other debris, technology not fully ready
Soybeans	Biodiesel	49	High	Low–medium	Medium	Medium–low	Technology ready, reduces food supply
Rapeseed, canola	Biodiesel	37	High	Medium	Medium	Medium–low	Technology ready, reduces food supply
Algae	Biodiesel	–183	Medium	Low	Low	High	Potential for huge production levels, technology not ready

Figure 6.3
An evaluation of different types of biofuel

Criticism

Criticism of biofuels is often centred on issues of land-take, that is, taking agricultural land that was used for food production and switching to crops for bioenergy. This is especially heightened in developing countries where there are already food shortages and malnourishment. Large areas of land clearance for biofuels are also linked to deforestation. The World Bank is just one of the organisations that has highlighted the 'fuel vs food' debate. It suggested in 2008 that '…the crop required to fill the tank of an SUV (Sports Utility Vehicle) with ethanol (240 kg maize for 100 litres of ethanol) could feed one person for a year'.

Linked problems include issues of soil erosion, biodiversity loss and nutrient leaching. Large-scale land conversion for biofuels may also push up global food prices in the medium to longer term.

Another concern is that of economics. Unless biofuels can be grown and processed into a useable fuel at a competitive price against oil, then the uptake of the technology will be slow.

Benefits

Biofuels are not without their merits. In comparison with traditional fossil fuels they have much lower emissions. But biofuels are not **carbon-neutral**, as it requires energy to grow the crops and then convert them into fuel. The amount of fuel used during this production (to power machinery, produce and apply fertilisers, transport crops, etc.) does have a large impact on the overall savings achieved by biofuels. However, supporters of biofuels still argue that they are substantially more environmentally friendly than their alternatives.

Biofuels also have the potential to reduce and stabilise the price of oil. Even countries such as the USA are keen on home-grown biofuels since they increase fuel security (see page 25 and *Case study 2*) and reduce dependence on imported oil.

The Overseas Development Institute has pointed to wider economic growth and increased employment opportunities, along with the positive effect on energy prices, as reasons to back biofuel production. In Brazil for instance, an estimated 750,000 people are employed in sugar cane and ethanol production.

Biodiesel could offer a more acceptable long-term solution as it uses simpler/existing technology and has lower transportation costs, alongside increased labour.

The future?

There is no doubt that the biofuels debate is complex, but biofuels do provide an enormous opportunity to address climate change, energy security and development. Decisions about how and where to plant biofuels are important, especially if we want to avoid generating new environmental and social problems. There are also new technological advances on the horizon, including the large-scale development of biogas, which may mean a larger proportion of electricity produced using biomass power. Meanwhile, many energy TNCs are experimenting with new biofuel sources, including algae, which have the potential to be more environmentally friendly.

Other renewables

BRINGING RENEWABLE ENERGY TO ISOLATED COMMUNITIES · Case study **16**

One of the big advantages of renewable energy is its ability to meet the energy needs of people who might be **off-grid**. Some 1.5 billion people worldwide still lack access to electricity, and approximately 2.6 billion are reliant on wood, straw, charcoal or animal dung for cooking their daily meals. In many rural areas of developing countries, connections to electric grids may take decades or may never happen. Today, there is a wide range of renewable alternatives to grid electricity and carbon-based fuels.

In even the remotest areas, renewable energy sources such as household PV (photovoltaic) systems, micro-hydro-powered mini-grids and solar pumps can provide some of the basic necessities and comforts, including electric light, communications and heating/cooling. More recently, there have been encouraging developments with biofuel-based generating systems. The renewable sources that can be established in rural areas are summarised in Table 6.2.

Statistics on renewable energy use in rural areas of developing countries are not easy to obtain. It is therefore difficult to detail the progress of renewable energy in off-grid areas for all developing countries. Examples of schemes that have been adopted and promoted in remote areas of the developing world include the following.

Solar power in Bangladesh
In Bangladesh, the power grid reaches only about one-third of the rural population. In the early 2000s, the government and donors established a rural energy fund that has enabled a group of companies to install about half a million home solar systems. A key

Table 6.2 Transitions to renewable energy in rural (off-grid) areas

Rural energy service	Existing off-grid rural energy sources	Examples of new and renewable energy sources
Lighting and other small electric needs (homes, schools, street lighting, telecoms, hand tools, vaccine storage)	Candles, paraffin, batteries, central battery recharging by carting batteries to grid	Hydro-power (pico-, micro- and small-scale) Biogas from household-scale digester Small-scale biomass gasifier with gas engine Village-scale mini-grids and solar/wind hybrid systems Solar home systems
Communications (televisions, radios, cell phones)	Dry cell batteries, central battery recharging by carting batteries to grid	Hydro-power (pico-, micro- and small-scale) Biogas from household-scale digester Small-scale biomass gasifier with gas engine Village-scale mini-grids and solar/wind hybrid systems Solar home systems
Cooking (homes, commercial stoves and ovens)	Burning wood, dung or straw in open fires at 15% efficiency	Improved cooking stoves (fuel wood, crop wastes) with efficiencies above 25% Biogas from household-scale digester Solar cookers
Heating and cooling (crop drying and other agricultural processing, hot water)	Mostly open fires from wood, dung and straw	Improved heating stoves Biogas from small and medium digesters Solar crop dryers Solar water heaters Ice making for food preservation Fans from small-grid renewable system
Motive power (small industry)	Diesel engines and generators	Small electricity grid systems from micro-hydro, gasifiers, direct combustion and large biodigesters
Water pumping (agriculture and drinking)	Diesel pumps and generators	Mechanical wind pumps Solar PV pumps Small electricity grid systems from micro-hydro, gasifiers, direct combustion and large biodigesters

part of this programme has been to ensure that the systems meet high quality standards and to provide guarantees for the technology and after-sales service.

Improved cookstoves

In rural areas of developing countries, most energy used for cooking is in the form of wood, straw and animal dung. The cooking stoves used are often quite primitive and have poor combustion efficiency, wasting power and causing health problems as people inhale the fumes. Today, a new generation of improved biomass stoves is being manufactured in factories and local workshops. The stoves are made of durable materials that will last for 5–10 years or even longer, and many are sold with guarantees. Such stoves improve the energy efficiency of cooking and lower indoor air pollution. Longer-term they are much cheaper to run and operate.

The World Health Organization and the United Nations recently surveyed 140 countries with a combined population of 3 billion people who rely on solid fuels such as wood, straw, dung and coal for cooking. The study found that approximately 30% are now using improved stoves. This amounts to about 166 million households, including 116 million in China and more than 13 million in the rest of east Asia, 20 million in south Asia, 7 million in sub-Saharan Africa, and over 8 million in Latin America.

Irrigation in India

In India today there are approximately 7,000 solar-powered pumps for irrigation. These systems are not as numerous as the technologies used for lighting, communication or cooking, but they can be important for increasing income in rural areas.

Small-scale hydro in China

In the early days of China's rural electrification programme, small-scale hydro systems were promoted to provide energy self-sufficiency to isolated local communities. Today, as the country's electricity grid expands, many of the same hydro stations provide power to the grid system. As of 2007, some 50 GW of small-scale hydro was installed in China, of which only about 3 GW was not connected to the existing grid system.

11

Using case studies

Question

Explain how the above schemes are examples of both intermediate and appropriate technology.

Guidance

Use www.practicalaction.org.uk to support your answer. Look for low cost, simplicity of use and maintenance, and ability to improve quality of life for developing countries.

AN ASSESSMENT OF GEOTHERMAL ENERGY IN ICELAND

Case study **17**

Iceland lies on the mid-Atlantic ridge, a constructive plate boundary. This means that the island is especially well placed to use geothermal energy. Rock temperatures are highest next to volcanic fractures. In some parts of Iceland, the temperatures of the rocks can be 100°C even at sea level. The temperature increases with depth and at 2 km down temperatures are more typically 300°C.

Power generation

In previous centuries geothermal heat was only used by households bathing and washing clothes. In 1930 the Reykjavik District Heating System was introduced, bringing hot water to the homes within the city from a series of nearby boreholes. In 1990 a power station was opened to the east of Reykjavik at Nesjavellir. This uses the hot water to generate electricity. It supples 60 million m^3 of hot water annually (around 1,800 litres per second) and 120 MW of electricity to the Greater Reykjavík Area. There are 21 boreholes used by the power station, each with a depth ranging between 1,000 m and 2,000 m. Temperatures of up to 380°C have been measured at the bottom of some of the boreholes.

Generated electricity is fed to Reykjavik directly via overhead and underground transmission lines. The transfer of hot water, however, is more complex. It is initially pumped through an insulated 90 cm pipe to a rock ridge at over 400 m. There it is stored in a series of large insulated tanks ready to be distributed towards the capital via a gravity-fed system (see Figure 6.5). Incredibly, the temperature drop along the course of the 30 km pipeline towards Reykjavik is only 2°C from start to finish. As a measure of the pipeline's insulation, where it lies above ground snow does not even melt on it.

Figure 6.4 The geothermal Nesjavellir power station, Iceland

Figure 6.5 Iceland's conduit supply system

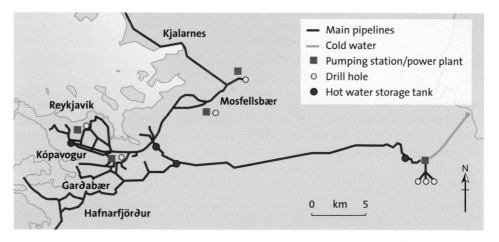

Main pipelines
Cold water
■ Pumping station/power plant
○ Drill hole
● Hot water storage tank

Kjalarnes
Reykjavik
Mosfellsbær
Kópavogur
Garðabær
Hafnarfjörður

0   km   5

N

There are four other major geothermal power plants in Iceland, which produce approximately 24% (in 2009) of the nation's energy. Geothermal energy meets the heating and hot water requirements of approximately 87% of all households in Iceland (nearly all of whom are in the capital). Apart from geothermal energy, 75% of the nation's electricity is generated by hydroelectric power, and only 0.1% from fossil fuels.

Environmental concerns

Geothermal energy, however, is not without its environmental concerns. The steam from geothermal areas contains dissolved gases that are released into the atmosphere.

The main pollutants are the greenhouse gas carbon dioxide and hydrogen sulphide (H_2S) which is toxic. About 7,500 tonnes of carbon dioxide annually are released into the air from the Nesjavellir plant — this is about one-sixth of the carbon dioxide that a relatively clean natural-gas-fuelled power plant produces. Emissions of hydrogen sulphide, although lower than carbon dioxide (e.g. 1,700 tonnes from Nesjavellir), can be converted into other chemicals. There is, however, no evidence of this causing damage in Iceland as research findings indicate that the hydrogen sulphide forms a harmless salt.

Overall, geothermal power is cost effective, reliable, sustainable and environmentally friendly, but has historically been limited to areas near tectonic plate boundaries. Worldwide, geothermal energy is generated in over 20 countries. The USA is the world's largest producer, with the largest geothermal development in the world at The Geysers, north of San Francisco in California.

Recent technological advances have dramatically expanded the range and size of viable geothermal resources, especially for applications such as localised/community and home heating. Engineered geothermal systems (EGS) will work in areas of the world that are not volcanically active. The process involves drilling thousands of metres underground but aside from that the design is the same as for natural steam or hot-water reservoirs. The big advantage of this is that it can work virtually anywhere in the world, but development of this new source of geothermal energy will require both technical and economic hurdles to be overcome. At the moment only a few EGS systems exist worldwide, including a pilot plant in Soultz, France, and a small commercial system in Landau, Germany.

12 Using case studies

Question
Explain why geothermal energy is unlikely to make a significant contribution to global energy supplies.

Guidance
Highly localised for geological reasons. EGS may allow its development in other areas, for example some locations in the UK are being considered (e.g. Cornwall). Mainly used for hot water supplies.

Tidal power is another example of a locally significant energy source, as the number of appropriate tidal locations around the world is limited. However, where there is a suitable resource, schemes have the potential to make a large-scale contribution, unlike many other renewables.

WEIGHING UP THE SEVERN BARRAGE TIDAL SCHEME

Case study **18**

The potential for power generation with the Severn Barrage tidal scheme is massive — equivalent to around three new large nuclear power stations. However, the plan is guaranteed to permanently disrupt the wetlands and mud flats behind the barrage, reducing the range between high and low tides by 50%. This is another example of a potential renewable source that is surrounded in controversy and disagreement between a range of **players** and stakeholders.

The Severn Estuary

The Severn Estuary lies on the southwest coast of the UK between Wales and England at the mouth of the Severn, Wye and Avon rivers. Its potential as a source of tidal power has been the subject of studies for well over 60 years. The estuary has the second highest tidal range in the world after the Bay of Fundy, Canada, with spring tides sometimes exceeding 14 m.

The entire area of the Bristol Channel and the Severn Estuary is recognised ecologically as extremely important. The estuary has been made a Special Protection Area (SPA) and designated a RAMSAR site. The area contains:

- 157 Sites of Special Scientific Interest (SSSIs)
- five National Nature Reserves (NNRs)
- two SPAs (Special Protected Areas)
- 13 of Britain's Heritage Coasts

Additionally, many coastal areas within the locality are designated as Areas of Outstanding Natural Beauty (AONBs).

Pros and cons

In the last 5 years the UK government has commissioned a number of large-scale reviews to examine the pros and cons of the Severn tidal resource, not least because of the very big costs and environmental concerns. Although a large number of different locations for the barrages were initially proposed, the Department for Energy and Climate Change (DECC) review has now shortlisted three main barrage schemes and two tidal lagoons (see Figure 6.6 and Table 6.3).

Table 6.3 *The Severn Barrage tidal scheme*

Figure 6.6 reference	Name of scheme	Description	Maximum power output
1	Beachley Barrage	The smallest barrage on the proposed shortlist, just above the Wye River	625 MW
2	Shoots Barrage	A small scheme upstream of the Cardiff-Weston Barrage, which could generate a similar amount of energy to a large coal-fired power station	1.05 GW
3	Fleming Lagoon	A non-barrage scheme on the Welsh shore of the estuary between Newport and the Severn road crossings	1.36 GW
4	The Cardiff-Weston Barrage	A 16 km barrage crossing the Severn Estuary from Brean Down, near Burnham, to Lavernock Point, near Cardiff, which could generate nearly 5% of the UK's electricity	8.6 GW
5	Bridgwater Bay Lagoon	A proposal which would impound a section of the estuary on the coast between Hinkley Point and Brean Down	1.36 GW

Environmentalists are more in favour of the lower-impact concept of tidal lagoons, whereby man-made lagoons in the estuary would fill and drain through turbines. Another advantage with these is that the lagoons could be sub-divided, so power would be generated at more states of the tide than with a barrage, albeit with lower peak

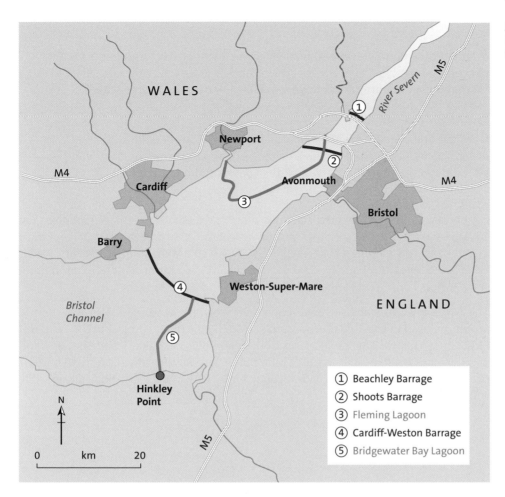

Figure 6.6
The Severn Barrage tidal scheme

① Beachley Barrage
② Shoots Barrage
③ Fleming Lagoon
④ Cardiff-Weston Barrage
⑤ Bridgewater Bay Lagoon

output, giving economic advantages to set against the higher construction cost of longer barriers.

Might it happen in the future?

The Severn scheme is immensely controversial. On the one hand it offers a long-lasting, low-carbon renewable energy source and would create tens of thousands of jobs, but on the other hand the potential environmental impacts may be overwhelming and not fully understood. The scheme's future will depend not on the various pressure groups for and against it, but on commitment from a future government to at least part-finance the scheme, which could cost in the region of £20 billion.

Critics argue that there are better ways to generate clean power on a large scale. Research from the Electricity Policy Research Group at Cambridge University argues that the same amount of electricity could be generated cleanly at two-thirds of the price by building three new nuclear power stations. The report suggests that a barrage would be 'ludicrously expensive' and thinks it is likely to be shelved for the foreseeable future as the government will not underwrite such a large infrastructure project at a time of spending cuts.

Questions

1   Design a poster or PowerPoint presentation in which you evaluate the arguments for and against the Severn Barrage.
2   Do you agree that there are better ways to generate clean power on a large scale?

Guidance

1   Look at both positive and negative environmental and economic considerations.
2   Green, clean alternatives include building three to four nuclear power stations, or around ten new gas-fired stations, or even four large coal-fired power stations, retro-fitted for CCS.

In contrast to large-scale tidal schemes there have also been many technological advances with small-scale solar power installations, which a have huge future in tropical developing countries (see **www.solarindiaonline.com** or **www. barefootcollege.org**). Equally, wave power, which is so far at a more experimental stage, offers considerable potential for small island communities.

The UK is a world leader in wave power, for example with the LIMPET system, a prototype used in Isla, and more recently used in a £4 billion project in the Pentland Firth and Orkneys. This project uses both Pelamis ('sea snakes') and SeaGens and with ten sites could power up to 750,000 homes, about the same as a small nuclear power station. The question is, how economically viable and technologically feasible will these projects be?

The future energy challenge

Global primary energy demand is forecast to increase by 40% or more by 2030 compared with 2010 (see Figure 7.1). The majority of this increase will be in the non-OECD markets. Although renewables will grow quickly, this is from a relatively low base and oil and gas will remain indispensable to energy supply between now and the middle of this century. It is likely that electricity not supplied from renewables will have to be provided by burning of coal and natural gas. It is the current glut of gas, not the surge in renewables, that is presently threatening the place of coal as an energy provider.

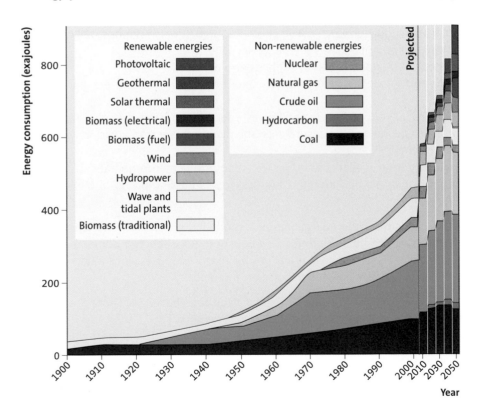

Figure 7.1
Worldwide energy consumption, 1900–2050

Energy supplies are therefore in a transitional state. Future increases in the demand for energy are unlikely to be even across the globe. China and India are expected to have the largest rises, while countries in western Europe may actually reduce their demands as a result of improved energy efficiency measures set against a backdrop of increasing fuel costs. By 2035, developing countries could be consuming nearly 10,000 million tonnes of oil equivalent (Mtoe) annually, compared with some 6,000 Mtoe in the world's wealthiest countries. As Part 2 showed (see Figure 2.2 on page 29), there is a strong correlation between energy consumption and the level of a country's development.

A big question for the global community is how it can grow economically without placing the environment at risk. More than 1.5 billion people (about one in four people on the planet) — have no access to modern energy to light and heat their homes. Universal access to clean, modern energy is considered a right for every human.

Energy security and the energy gap

Many nations are worried about energy security — affordable energy is vital to the functioning of a country, particularly for its economy and the quality of life of its inhabitants. Rapid increases in the costs of fuel, for example, have a big impact on logistics and supply chains, so volatility in energy supply is undesirable. There are a number of interlinked risks to energy security:

- **Physical** — the exhaustion of supplies and reserves.
- **Economic** — sudden rises in the costs of energy; this may affect global interest rates.
- **Geopolitical** — political instability in energy-producing regions, disputes and conflicts over ownership/sovereignty and disputes over transmission, for example gas pipelines.
- **Environmental** — worries and protests about the environmental damage caused by the exploitation of energy resources, for example the Gulf of Mexico oil spill in 2010.

Figure 7.2 illustrates the energy gap as a modelled prediction. The orange section on the graph represents the difference between global energy demand and global energy supply. The energy gap is a worry for many countries since it decreases their energy security and increases reliance on imports.

So, managing the future energy mix is a daunting challenge. There are two main pathways open to governments and decision makers:

- Application of various 'carrot and stick' measures, such as emission controls, carbon trading and green taxation to encourage a reduction in energy consumption and an increase in energy efficiency (see Figure 7.3). Liberalised energy markets also fall within this category (see page 101).
- Developing new and radical technologies that are sustainable, future-proof and bring energy security. These are considered in more detail later in Part 7.

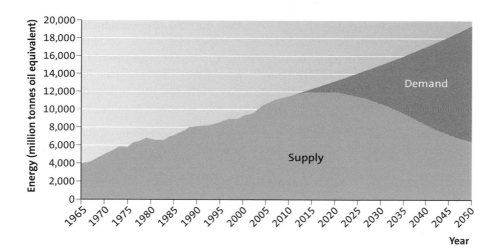

Figure 7.2
A visualisation of the energy gap

Emission controls

The Kyoto Protocol (1997) proposed emission controls at the international level for the first time. It came into force in 2008 and includes limited controls for some of the emerging economies. The aim is to reduce greenhouse emissions by an average of 5% (1990 levels) by 2012. The refusal of the USA to ratify the protocol weakened its standing

Emissions trading

Countries that have made greater reductions in their carbon emissions (than set out by Kyoto) can sell their surplus savings to countries that have exceeded their targets. This has created a new commodity market, where credits can be traded. The system means that countries can go on polluting as long as they can pay

Green taxes

Green taxes have been introduced in some countries to try to reduce resource consumption and increase recycling. In the UK large, more polluting cars pay more vehicle excise duty (car tax) than super-efficient hybrids and smaller cars. There are also green taxes on air travel and petrol/diesel

Figure 7.3 *'Carrot and stick' measures*

Can technology secure future energy supplies?

Technology has the potential to ameliorate the impacts of growing demands and falling energy outputs. Technology can offer solutions to our energy 'fix' in a number of ways:

- Discovery of **new locations of energy**, for example new oilfields and tar sands (see Part 4). Technology will improve the feasibility and recoverability of such sources, especially when accompanied by an increase in the price of energy to make exploration more cost effective.
- Discovery of **new types and sources of fuel or energy**, including renewables such as wave power.
- Assistance in the **large-scale storage** of energy. This is one of the key objectives, to be able to 'smooth' the flow of energy from renewable technologies.
- Delivering **energy more efficiently**, for example smart grids/dynamic demand (see *Case study 20*).

- Help in the **removal and/or storage of carbon dioxide**, for example through new carbon sequestration technologies (see *Case study 21*).
- **Improving energy efficiency in vehicles and in our homes** (see *Case study 19*), thereby reducing overall demand and consumption.
- **Improving efficiency of resource extraction methods and conversion to end-use form**, for example using new turbine technology when generating electricity.

New energy technology brings a number of linked economic benefits, most obviously more employment in the energy sector. The US government, for example, agreed that between 2009 and 2019 it would invest over $150 billion in a 'clean energy economy'. This would involve the development of large-scale renewables, new fuel infrastructures and a new digital electricity grid. It is hoped that these investments will create 5 million new 'green' jobs in the private sector.

There will be other knock-on effects for countries championing new energy technology — one of which is the ability to sell knowledge and specialist products to emerging economies without a mature energy technology sector. Established high-income countries such as the UK and Germany see this as an opportunity for generating new foreign markets and capital flows. Equally, the emerging super-powers such as China and India have seized this opportunity to market lower-cost solar and wind technology around the world.

Case study 19 DEVELOPING ECO-HOMES FOR FUTURE ENERGY SAVING

The homes and buildings within the EU use the equivalent of 6 million barrels of oil a day to heat them — a figure that could be halved if they were better insulated. This would cut about €270 million from the EU's total energy bills and 460 million tonnes from its annual carbon dioxide emissions, easily meeting its Kyoto commitments. The USA could potentially save $1.2 trillion by 2020 if it improved energy efficiency.

Until now most of the effort to make housing more energy efficient has focused on new builds. In 2007, the UK government made an ambitious promise to make all new houses carbon-neutral by 2016. Germany has even tighter building regulations for energy efficiency: the PassivHaus standard requires homes to use just 10% of the energy of a typical house.

Figure 7.4 shows some of the possible ways of making houses more eco-friendly. Payback time, annual carbon dioxide savings and approximate costs are also indicated. According to this the best-value energy-saving measures are cavity wall insulation and loft insulation as they have short pay-back times.

Loans for energy savings

The biggest barrier for homeowners is the cost of retro-fitting energy-saving measures. But there are models in place to help finance some eco-savings. In 2007, Berkeley, California, launched itself as a Sustainable Energy Financing District, lending funds to homeowners wanting to install solar-thermal or photovoltaic panels on their roofs. The loan is repaid through a local taxation scheme, but this is more than offset by the reduction in energy bills. An important feature of this is that the debt stays with the house, so if the home is sold the solar panels are transferred to the new owners, along with the remainder of the loan. A number of other US states are likely to adopt the Berkeley model in the next few years.

Ways to save	Payback time (years)	Installed costs (£)	Annual carbon dioxide saving (kg)
Loft insulation	2	250	800
Condensing boiler	12	3,000	1,260
Small mast-mounted wind turbine	40	15,000	2,600
Cavity wall insulation	2	250	610
Draught-proofing	8	200	130
Ground-source heat pump	12 (electricity) 50 (gas)	10,000	6,000 (electricity) 0 (gas)
Solid wall insulation (external)	30	13,000	2,100
Solid wall insulation (internal)	18	7,000	2,000
Double glazing	35	5,000	720
Solar water heater	50 (electricity) 80 (gas)	4,000	580 (electricity) 260 (gas)
Solar electricity	~55	11,000	1,000

Figure 7.4
How to make your house more eco-friendly

The UK government too has recently announced a number of long-term loans for energy-saving retrofits. As with the US model, the loan is secured with the house so the debt is transferred should the house be sold to a new owner. This UK scheme is working alongside the 'feed-in tariff' (FIT) scheme whereby homeowners are paid to generate their own renewable electricity. Photovoltaics fitted to an existing property, for instance, will attract 41p per kWh per year, which is guaranteed for 25 years. A typical 2 kW installation generating 2,000 kWh per year means the householder receives an annual payment of £820, which is in addition to the savings on electricity generated. It is hoped that FITs will encourage take-up of renewable resources by individual households and community groups.

The technological fix needs to be supported by an attitudinal fix whereby householders become champions of energy conservation.

Figure 7.5
(a) Retro-fitted photovoltaics in Dorset, made financially viable because of the FIT scheme (b) An energy output display panel

David Holmes

DYNAMIC DEMAND AND SMART GRID SYSTEMS

One of the key problems with renewable sources is their intermittent availability (see page 79). In many instances, electricity can only be generated when the wind is blowing or the sun is shining. Critics claim that this is a real barrier to wholesale renewable adoption, especially for electricity generation, since supply will not be able to match demand.

When there is an increase in demand on the grid or supply drops (as might happen if the wind stops blowing at a large wind farm for instance) the frequency of the grid will dip below acceptable limits. In Europe the frequency should be about 50 Hz (hertz), i.e. 50 cycles per second. If the frequency drops to below 48.8 Hz then grid operators must shed some of the load, and parts of the country are disconnected from the grid and black out. However, new technology could smooth the mismatch between supply and demand.

Dynamic demand

Dynamic demand relies on a network that is controlled by individual appliances within households, rather than a central control room. It works on the assumption that as load on the grid increases, the frequency drops slightly. In a pilot study in Italy, 'intelligent' domestic fridges are being used to control the electrical loading on the grid. As the frequency drops, a built in controller checks the temperature of the fridge and calculates how long it can stay chilled without drawing any power. It then switches the fridge off for however long is safe, thereby reducing load at that moment on the overall grid. If this technology (only costing about £4 per fridge) were fitted to enough fridges and other appliances (e.g. air conditioning units and freezers) it could go a long way to smoothing out the fluctuations caused by the intermittent nature of renewable energy supplies.

Smart grid systems

In a contrasting approach, known as a smart grid system (Figure 7.6), an operating system run by the electricity company uses two-way communication between the grid and the householder's appliances. Using information fed by the appliances, combined with predictions of renewable power based on short-term weather forecasts, the operating system can balance demand and supply by shutting-off non-essential appliances within the house. By providing homes with smart meters to monitor electricity use, such systems can help smooth out demand by also altering the price per unit of electricity. When there is a surplus it is cheaper, and when there is a shortage it becomes more expensive. Consumer behaviour can be influenced in this way, for example by encouraging people to set their washing machines to run at cheaper off-peak times.

If electric cars become widely used in the future it may be possible to incorporate these into the smart grid system of a home. When there is a low demand the car batteries could be charged ready to act as a store. During times of heavy loading or sudden demand, electrical energy from the batteries could be fed back out, thereby increasing the amount of available electricity. If this were done in thousands of households it would have a significant impact on the total energy available to supply the grid.

Contemporary Case Studies

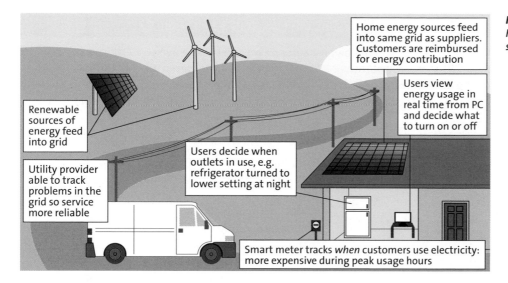

Figure 7.6
How a smart grid system operates

Pros and cons

Dynamic demand and smart grid systems are not without their problems. One particular issue is the agreement of common standards for the technology required for the dynamic demand plug-in modules, or for the smart grids. This would have to be a globally agreed solution. In addition white goods (e.g. fridges) are changed relatively infrequently by households, so it would take a long time for all consumers to replace their existing fridges with ones able to respond to dynamic demand. Electricity companies are also worried about the cost of smart-meter installation, which is estimated to be between £250 and £450 per household. Smart meters will, however, do away for the need to take individual meter readings and will allow for much more accurate billing.

Smart metering and dynamic demand are likely to be welcomed by many consumers, especially against the backdrop of rising electricity prices. If people have better control over their electricity use then this is likely to reduce demand and at least offset the rising long-term costs. A recent study in the Netherlands of smart grids revealed an interesting pattern of costs and benefits for different stakeholders — see Figure 7.7.

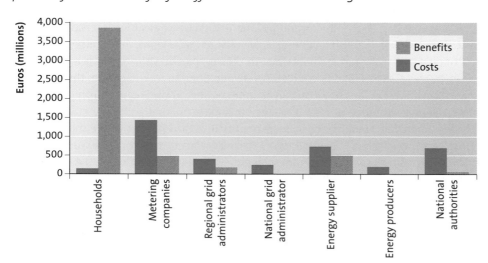

Figure 7.7
Overview of costs and benefits of smart grids for different stakeholders

China is now the world's biggest carbon emitter, yet the average Chinese person uses much less energy than their US or European counterparts. The reason is obvious — China is a vast country with the world's largest population. It is also a rapidly emerging economy so the demand for energy is rising at an unprecedented level.

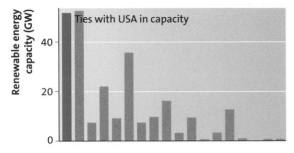

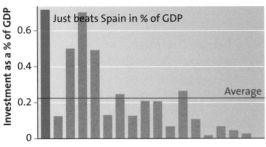

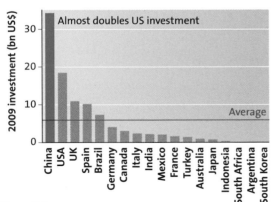

Figure 7.8
Data on spending on renewables for China and other nations

Reducing emissions

But since 2006 China has put together impressive programmes to fight global warming. As a rapidly developing nation it needs to build power stations, road networks, ports and airports as well as industrial infrastructure. So China's emissions reduction target cannot happen in the next few years while it is still trying to achieve its development goals. Despite this, the government has ambitious targets to reduce energy intensity — emissions per unit of GDP.

In China's 5-year plan (2011–15) it is proposing deep cuts in energy intensity, as well as increases in zero- and low-carbon energy production, and plans to create more carbon sink forests. The pace and scale of expected change are impressive (especially when compared with European promises and US inaction). For example, by 2020 China's energy mix will include 15% non-fossil fuels, up from just 7% in 2009. Hundreds of billions of dollars are being invested in wind, solar, micro-hydro, biomass and nuclear power. In 2009 China invested $35 billion into renewables; the figure for the USA was only $18 billion (see Figure 7.8).

Sharing knowledge

Various partners and international organisations are working to share knowledge and technology with China so that it can achieve its aspirations as a lower-carbon economy. The British Geological Survey (BGS) and the China University of Petroleum in Beijing have been working in close partnership with Herriot Watt University, BP and Shell to investigate the potential of carbon dioxide sequestration and geological storage.

Measures to improve energy efficiency include new fuel economy standards for vehicles, improved building codes, investments in ultra-efficient conventional power plants with new grids, together with financial incentives for renewable power. Innovations in areas such as thin-film solar panels, battery technologies, hybrid vehicles and carbon capture are also being used.

Impacts of climate change in China

China has a particular interest in trying to act against climate change. The glaciers in Tibet for instance are shrinking fast and this will affect China's water supply. It is also likely that climate change will bring more extreme weather to China in the form of heatwaves, intense rainfall and flooding, as well as droughts. There may also be shifts in regional weather, causing changes to agricultural patterns and productivity. Sea-level rise would impact on China's delta regions, which form today's major population and economic centres. Moreover, China is conscious of the impact of high levels of air pollution (caused by the present day reliance on coal) on people's health.

It is too early to say when China's carbon emissions will peak. Various scenarios are proposed (see Figure 7.9), but many experts predict a peak somewhere between 2030 and 2040. After this, emissions should start to fall, and a low-carbon scenario indicates that this will be rapid. With China's track record of effective action, it could end up as a model low-carbon economy.

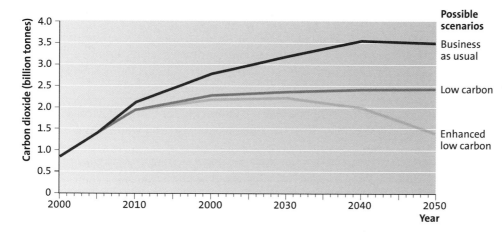

Figure 7.9
China's emissions target: carbon dioxide output from fossil fuels, measured in billions of tonnes of pure carbon

14 Question

How feasible do you think this aim is? What factors are likely to impede progress?

Guidance

The constant drive for economic growth and rising living standards may mean the reliance on coal is essential for the immediate future. Think about cost and also short-term versus long-term gains.

Liberalised energy markets: will they work?

Within Europe, industries and private households are in theory able to freely choose their energy supplier following the coming into force of EU directives on competitiveness in 2004 and 2007 (see *Case study 4*). This is at the core of a liberalised energy system. In the UK, liberalised energy was adopted by the Conservative government in the 1990s — supposedly offering lower prices for consumers through a competitive market. Certainly, since market liberalisation, British customers have paid less for gas and electricity than some of their European counterparts. On the downside, long-term planning of electricity generation to overcome a potential electricity gap (see *Case study 7*) has been hampered considerably.

Market liberalisation is a complex issue. One of the key issues in the debate between liberalisation of natural gas markets and security of supply in both the EU and the USA relates to the impact on long-term contracts and implications for price volatility and levels of investment. Liberalisation supporters argue that even if long-term contracts are banned, markets can still deliver on security of supply as long as alternative trading arrangements exist. Opponents argue, however, that without long-term contracts between exporters and wholesalers, increasing price risk would result in less and/or delayed infrastructure investment, generating security of supply problems.

Within the EU there are also complex and controversial links between the production and distribution arms of large integrated energy firms such as France's EDF and Germany's E.ON. In effect this means that these organisations increasingly have an energy monopoly, thus obscuring the market principle.

As a result of potential energy shortfalls, outsourcing of energy development has been proposed as a solution — just the same as for food and water supplies.

Case study 22 SUNSHINE SUPERPOWER: OUTSOURCING THE SAHARA

Figure 7.10
World insolation

Figure 7.10 shows the global solar energy potential of the world. The highest concentrations of solar energy are found in parts of Africa around the equator — this is a huge solar powerhouse that has the potential to be tapped for energy.

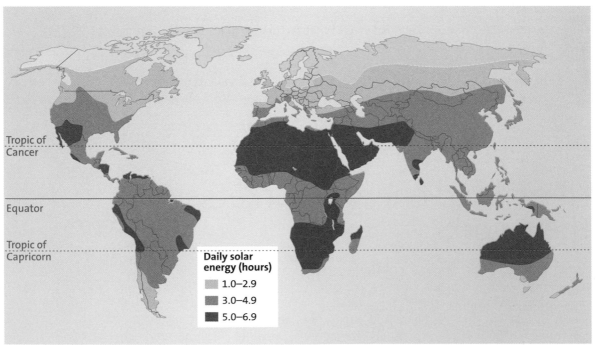

Contemporary Case Studies

Desertec

A large project called Desertec is currently being evaluated to see whether solar energy can be collected over Africa and converted into electricity for use in European homes and businesses. It is claimed that the project, headed by 20 major German companies, could produce 15% of Europe's electricity needs by 2030. €400 billion would be needed to build a raft of solar thermal power plants in north Africa. Peak power is projected at 100 GW — roughly equivalent to 100 coal-fired power stations.

Under the Desertec proposal, concentrating solar power systems, PV systems and wind parks would be located on 17,000 km² of the Sahara Desert. Electricity would reach Europe via 20 high-voltage direct current power lines, which would keep energy losses below 10%. Trans-Mediterranean links would cross from Morocco to Spain across the Strait of Gibraltar, from Algeria to France (via the Balearics), from Tunisia to Italy, from Libya to Greece and from Egypt to Turkey via Cyprus (see Figure 7.11). The proposal would integrate and manage supply of renewable energy from all over Europe to try to ensure a steady flow of electricity to end-users.

DESERTEC/www.desertec.org/

Figure 7.11
The Desertec project

Concerns

Desertec is an ambitious and massively expensive project. Critics say that it could make Europe's energy supply a hostage to politically unstable countries. Others argue that it is exploiting sunshine from Africa that should be used by Africa. It may also represent a poor investment compared with insulating Europe's homes and businesses or helping to cover the roofs of people's houses with PV.

Another concern is the amount of water required to operate a solar thermal power plant. Water is used a bit like it is in a conventional power plant to generate steam to drive turbines. A similar designed plant in the Mojave Desert consumes around 3,000 litres of water for every MWh of electricity. The Saharan plan would need equal amounts of water. This is very problematic in the desert, and although water can be abstracted from aquifers, this is a non-renewable resource. As with many techno-fix solutions, the economic and political costs of the grand design may in the end lead to it remaining a vision of the future — so much depends on the supply/demand equation.

Using case studies

15

Question
Write an appraisal of the future of solar power globally.

Guidance
Use Figure 7.10 to assess the most likely locations for development and then look at the various technological advances for both large- and small-scale use.

Energy futures: final thoughts

This book has pointed out that fossil fuels will become more depleted and less economic to use with time, as well as environmentally unacceptable. Peak oil may be on the horizon and most scientific research points to carbon-linked climate change occurring, which is having some political impacts across the globe.

The world's population will continue to rise and may peak at 9–10 billion by 2050 — this will be combined with big increases in demand for fuel from emerging economies, particularly the **BRICs**. As more countries seek to improve their peoples' quality of life, there will be an increased burden on energy, currently from fossil fuels. We have already seen that there is a strong positive correlation between HDI ranking and energy consumption per capita.

Future scenarios

Much work has been done to model different future energy scenarios by national governments and international agencies such as IEA, as well as energy TNCs. Essentially they look at business-as-usual, sustainable pathways and radical action options. In the UK, for example, **OFGEM** has proposed four different energy scenarios (see Table 7.1).

Each of the different scenarios is based on a series of assumptions. For example, the 'dash for energy' might be thought of as 'business as usual', whereas 'green stimulus' is based on the idea that there is considerable investment in renewables and carbon storage as a way of stimulating the economy. 'Green transition' assumes a drop in energy demand due to technological efficiency and 'slow growth' supposes slow development of renewables and no extensions to nuclear power.

Table 7.1 OFGEM's future energy markets

	Green transition	Green stimulus	Dash for energy	Slow growth
Economic recovery	Rapid	Slow	Rapid	Slow
Environmental actions	Rapid: renewables targets met, investment in CCS	Rapid: renewables targets met, investment in CCS	Slow: renewables targets not met, limited CCS	Slow: renewables targets not met, no CCS
Gas demand	Falls: energy efficiency, renewables heat	Falls: recession, energy efficiency, renewables heat	Increases	Falls until 2012, recession, then increases
Electricity demand	Falls until 2015: energy efficiency. Increases longer term — electrification of heat, transport	Falls until 2015: energy efficiency. Increases longer term — electrification of heat, transport	Increases	Falls until 2012: recession, then increases
Supply of pipeline gas to EU	Medium	Low	Medium (high production, but diverted to East)	Low
Global LNG market	Tight in the medium term before falling back	Over-supplied	Tight	Over-supplied initially, becoming tighter
Commodity prices	Medium gas, high carbon, low coal	Low fuel prices, high carbon	High fuel prices, moderate carbon prices	Low fuel and carbon prices
Nuclear	Further extensions, strong new nuclear	Further extensions, strong new nuclear	No further extensions, new nuclear delayed	No further extensions, no new nuclear

Challenges and benefits

Britain's ability to meet its demand for energy is no doubt going to be tested over the next decade or so. Growing exposure to a potentially volatile global gas market and ageing power plants nearing the end of their life, along with the need to tackle climate change, are the central challenges that the country faces. These problems can be replicated in many other countries throughout the world; political procrastination and costs of replacing old infrastructure are the biggest barriers here. As individuals become more aware of energy challenges there will be increased pressure put on governments at a range of scales.

Yet there is a bright energy future too. Renewables will become more commonplace, especially in the medium term from 2030 onwards against the background of rising fossil fuel costs. There will be new jobs and careers, and perhaps a growing acceptance that fossil fuels are simply 'unacceptable to use' (rather like drink-driving has become nowadays). Larger communities are beginning to make the switch to renewables. Gussing, a town of 4,000 in eastern Austria, recently went 100% renewable in electricity production with an efficient 8MW biomass gasification plant fuelled by the region's oak trees. Freiburg, a city of 200,000 in southwest

Germany, has invested €43 million in PV installations over the past 20 years and has reduced its emissions by 25% compared with 1990s levels.

There are still big challenges in terms of managing growing world air transport, but smaller electric cars and hybrids will become cheaper and more accepted by consumers. Energy efficiency in our homes will be the norm, and new houses in many parts of the world should all be zero-carbon within 20–30 years. All that is needed is the investment to make it a reality.

Examination advice

Types of exam question

AS questions

At AS most questions are structured questions, usually involving data-response from interpreting source material, and then a series of short questions worth between 4 and 6 marks. Most of the AS command words are quite basic: describe, suggest reasons/explain, state, identify, outline, comment on or examine. You are responding to graphs or statistics that you may be asked to analyse. Most questions end with an opportunity for extended writing, which is also marked for quality of written communication.

Here are some typical examples of short, structured questions.

Knowledge-based questions

Describe the following terms:	
Stock resources	(2)
Flow resources	(2)
Secondary energy	(2)

Be guided by the mark scheme — you can get definitions worth 1, 2 or 3 marks and need to tailor your answer accordingly. (Learn all the key terms explained on pages 7–8 of this book.)

Answer

Secondary energy is power manufactured from primary resources (1) such as electricity from coal or petroleum from crude oil (1).

The examples enable you to achieve the extra mark.

Outline the contribution *stock resources* make to the UK's *primary energy mix*.	(4)

Here you are likely to be marked by point marking, so you need to identify *at least* four key statements to be safe and hit the points on the mark scheme.

Answer

Stock resources are finite (1). They include coal, oil and gas, i.e. fossil fuels (1) but they cannot be used without depleting the stock because their rate of formation, i.e. replacement, is so slow (1).

Note the question asks for contribution to primary energy mix *not* electricity and it asks you to outline the contribution, so some accurate numbers would be ideal.

Currently (in 2010) stock resources contribute 89% of the UK's energy mix (1) — coal 16%, oil 35% and natural gas 38% (1).

This amplification shows sound learning.

Skills questions: interpreting data

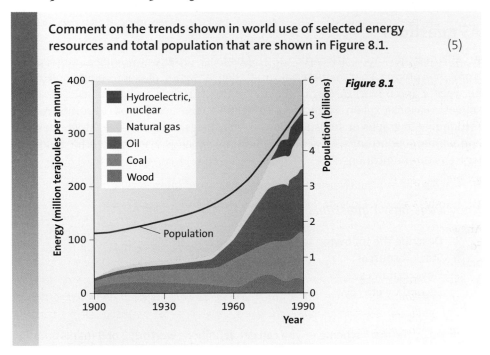

Comment on the trends shown in world use of selected energy resources and total population that are shown in Figure 8.1. (5)

Figure 8.1

Resource check: look at the units on both axes and the time frame. As this is an AS question you are not asked to link the two trends together, but the command word is comment on and there are 5 marks, so it would be sensible to talk about the link between the two trends. When describing graphs for AS always quote data to support your answer (as you did for GCSE) — you need a ruler to read off the figures accurately, so remember to take one to the exam.

Answer

The world population rises from just under 2 billion to 5.3 billion (1) over the 90 years between 1900 and 1990. The increase is especially marked after 1960 (1). This is reflected in the rising consumption of energy (link — 1), which actually increases at a greater rate (1) (not only more people but also increasing demands for both industry and domestic use) (1). In terms of individual trends wood fluctuates around 20 million terajoules, but also three fossil fuels increase dramatically in usage after 1960, especially oil and gas (1) — an upward trend except for a slight dip around 1973 during the world recession (1). Together they make up nearly 90% of total use in 1990 (1). Hydroelectricity and nuclear power only come on stream after around 1980 (1) with around 25 million terajoules in 1990 (1).

Note that there are around ten points made by the student in this very thorough answer, but it is succinct. Never make just five points for 5 marks, and always try to expand each point with an example.

Many questions are stepped, i.e. the questions get harder as you proceed through them. In the following example the mark tariff goes up to 10, so almost certainly the question will be marked on the basis of three levels.

To what extent (ct) are the trends (kw) shown in Figure 8.1 compatible (kw) with sustainable energy (kw) development? (10)

The first stage is to identify the key words (kw) as shown. The command term (ct) is *to what extent*, which means you have to consider how compatible the trends are. For a 10-mark question, a simple plan, as shown below, might help.

Plan

Define sustainable energy (environmental, economic, social) (1). Compatible (3) — use of HEP, nuclear — non greenhouse gas. Non compatible use of fossil fuels — finite, dirty, greenhouse gas (2).

The numbers are used to code your answer, so it reads in a logical order.

Answer
For 10 marks go for around 15–20 lines — be guided by the space in the answer booklet.
1. First paragraph — *define sustainable energy development. Issues of futurity (Brundtland) and eco-friendliness are paramount. Fossil fuels in particular are **finite** and polluting (especially coal) with nearly 90% reliance on fossil fuels. inevitable lack of compatibility.*
2. Second paragraph — *details of non-compatibility for three fossil fuels — possibilities of cleaner gas, even clean coal.*
3. Third paragraph — *development of hydroelectric (renewable, green) — dam issues? Nuclear non-polluting in greenhouse gas terms — other pollution issues.*

For this type of question case studies are not required, but examples (well-chosen, factual sentences) are useful to support arguments.

A2 questions

A2 questions always require extended writing. Many are structured into two parts — a data stimulus part (a), followed by an open-ended, essay-style part (b). Others might be just open-ended, essay-style questions. All are marked by using levels, and all include an assessment for QWC (quality of written communication). QWC includes developing a logical structure, using appropriate geographical terminology, and writing with good syntax, spelling and punctuation.

A third element in A2 assessment is **synopticity**, which is usually assessed by wide-ranging, often research-based essays (OCR) or an 'issues analysis' based on pre-release resources (Edexcel, AQA and WJEC). Synopticity is the ability to draw together all the knowledge, understanding and skills gained from the whole A-level course and to use it to answer designated synoptic questions or papers.

For standard A2 questions you usually have around 40–45 minutes, so the first process is planning (say 5 minutes overall), once you have made some notes on the **resource**.

Often the data stimulus part (a) is worth 10 marks, with the extended essay part (b) worth 15 marks, so the time has to be allocated accordingly.

Data stimulus questions

Figure 8.2 shows a typical A2 part (a) resource.

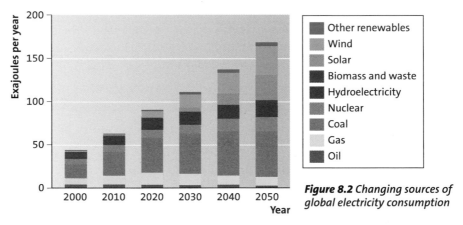

Figure 8.2 Changing sources of global electricity consumption

Study Figure 8.2 — Explain[4] the trends[3] in changing sources of global[2] electricity[1] consumption. (10)

The first task is to check out the resource[1]. It is a graph for electricity not energy consumption — this could affect the amount of oil, for example, as oil is largely used for transport rather than generating electricity. It is for global consumption[2]. It shows absolute amounts in exajoules, not per cent — if possible try to include statistics[3]. Note the command word[4] — there are no marks for slavishly describing the amounts used every 10 years — you have to take an overview.

Plan of trends

Oil — *steady* — *infinitesimal* — *reason: too valuable to use for electricity.*

Gas — *significant increase to 2030, slight decline by 2050. Supplies* — *onset of shale gas.*

Coal[B, D] — *in spite of dirty reputation* — *steady increase (reserved)* — *dominant* — *possible use of CCS. Issues of Kyoto targets. India/China (brief e.g.).*

Nuclear[A] — *steady increase — base load function — fills energy gap. Renaissance — 2010, no recent Chernobyl.*

Biomass waste — *negligible.*

Hydroelectricity[C] — *significant issues of dam construction (brief e.g.).*

Huge rise in **solar and wind**[C] *— key renewables — feasibility, viability, costs (brief e.g.). Need for non-greenhouse gas fuels. Limited by physical conditions.*

The issue is how to say this succinctly, interpreting key headlines, and bringing out key reasons:
- Need for security[A]
- Cost of supplies[B]
- Impact of Kyoto greenhouse gas targets[C]
- Rising demand, especially in India and China[D]

Try to get a *broad* overall coverage, so avoid long case studies, but try to support your explanation with brief examples as indicated — it all has to be done in only 20 minutes *max*.

Extended writing questions

For the extended writing questions (either part (b) or stand-alone essays with no resource), here are some examples. You may want to use one or two case studies, but they must be incorporated in the *overall* framework as opposed to being used descriptively. Note that all these essays have evaluative command words.

> 1 Assess the potential environmental, economic and political risks in exploiting new energy resources. (15–25)
> 2 Evaluate the view that the best way to achieve energy security is a balanced energy mix. (15–25)
> 3 With reference to two contrasting energy sources discuss the view that access to them is not evenly distributed. (15–25)

Issues to consider include:
- developing a sound structure, leading to a logical order
- assessing the focus required for the essay
- deciding on the use of case studies — all three titles have different requirements

Question 1 has a built-in structure for the three types of risk. A danger is failing to identify and justify what is a *new* energy resource. Legitimately it could include all the renewables and recyclables, possibly unconventional gas and oil (this would give you a chance to comment on shale gas or tar sands), and even nuclear. A possible plan could be:
- Political risk definition — could be linked to security, possible risk of very high costs and budget issues.
- Environmental risks — case studies of new green fuels and environmental issues.
- Economic risks — associated with economic viability — biomass possible case study.

Question 2 again has an inbuilt structure for defining energy mix and the concept of security (affordable, available, achievable — three As). Case studies could be related to a single country, for example the USA (*Case study 2*). Is a wide/balanced mix the best way — it may be very expensive, environmentally damaging etc. but obviously it is not necessarily all based on indigenous supplies. Include arguments for and against.

Question 3 is a classic two-case-study question, in this case requiring two *contrasting* sources. Clearly a commodity like oil or gas with its geopolitical emphasis is a must for one example. A contrasting resource could be widely distributed (e.g. coal) to argue against the statement in the question, or possibly hydroelectricity. Nuclear is a contrasting resource, which in theory is widely available but because of its military potential is restricted to rich countries and their friends (the case of Iran).

All three questions emphasise that there is no uniform approach in structuring, selecting or deciding on the length of case studies.

Developing essay-writing skills

Top tips

1 Get hold of past exam papers and mark schemes — many are available from exam board websites (for password access consult your teacher).
2 Use the exam practice sections in your standard textbooks. They have questions, mark schemes and quite often answers. The Philip Allan Updates series of *Student Unit Guides* includes student answers and examiner comments. There are also some new online products, some of which show you examples of good answers and also less good answers, with examiner comments. See, for example, the *Practise Every Question* series from Hodder Education.

3 Quite often your teachers will have supplies of past students' work. One way to understand what is required is to carry out a marking exercise to assess strengths and weaknesses.

4 Good essays require planning, especially when you are not working under timed conditions. You can plan titles in class as a group, or individually.

Possible plan formats

  4i Star diagrams — but always annotate them to establish a sequence.

  4ii Linear plans that include a short introduction, a main body and a conclusion.

   As is shown in the following example, the main body should be concept not case study driven.

 Assess the view that radical new approaches to meeting future energy needs are required to avoid environmental degradation.

Introduction

Definition of radical new approaches — a switch to clean fuels, nuclear and renewables and energy conservation to avoid problems of environmental degradation from greenhouse gases.

Main body

Techno fix: role of renewables

- How they avoid environmental degradation
- Controversy of nuclear avoids greenhouse gas issues but introduces new ones
- Development of CCS

HEP ✘ Three Gorges Dam — wind ✔ — solar ✔ — nuclear in UK, Chernobyl — CCS experiments

Attitudinal fix

- Energy conservation and efficiency

Case studies of energy conservation and efficiencies

Conclusion

Definitely yes

- Limits of techno fix
- Perhaps use of attitudinal fix
- Yes, some radical new approaches

4iii Spectrum plans — for some topics a simple spectrum diagram can be easy to use and will lead to a well-organised final product.

Evaluate the different environmental costs and benefits of a range of energy resources.

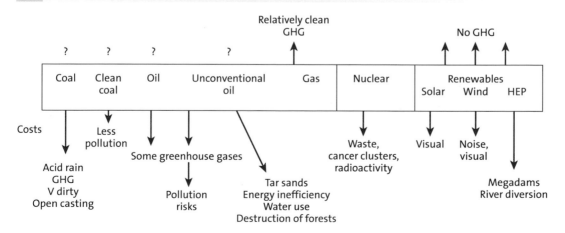

5 Before planning it is important that you deconstruct the title — it is vital to identify command terms (ct), key words (kw) and link words (lw), as shown overleaf.

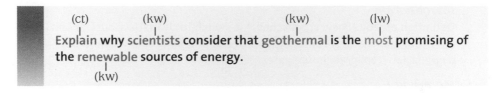

6 It is always worth working on your essay skills at A2 as this will build up your confidence. At A2 there is no real substitute for practice on a range of essay titles. Initially do not constrain the time you spend, but as you develop your skills practise under timed conditions. If you are chasing an A*, this is vital.

7 Some candidates never manage to write enough in the time allocated. One tip for saving time is to learn how to draw some well-organised diagrams, which can quickly be reproduced. Spider diagrams are a popular option (Figure 8.3).

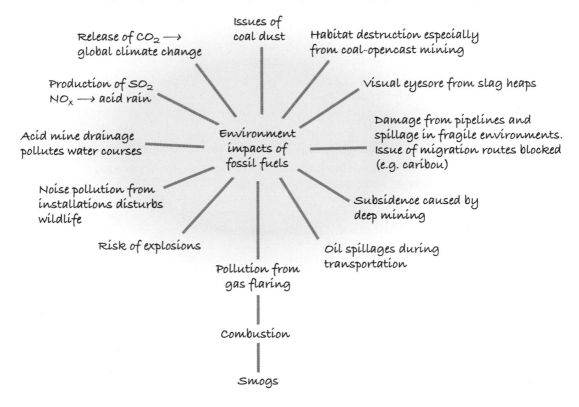

Figure 8.3 *Environmental impacts of fossil fuels*

Other good examples to learn are Hubbert's model of peak oil (see Figure 4.6, page 56) and the energy transition model (see Figure 2.3, page 30). Check out all the diagrams in this book and pick out some other suitable ones.

A further piece of advice is to always support your major arguments with brief case studies, as this helps to add depth.

Researching energy topics

Top tips

1 Use a range of books, articles and websites.

For **books** you should try to read beyond your textbook, using specialist energy texts such as this one (see 'Useful books' on page 9).

For **articles** you will find *Geo Fact Sheets* and *Geofile* useful as the articles are up to date and targeted at A-level students. Magazines such as *Geography Review*, *Geographical Magazine*, *National Geographic*, *New Scientist* and the *Economist* all have a wealth of up-to-date energy articles. Most schools have subscriptions for these magazines, some of which give password access to an associated website (e.g. *Geography Review*).

For **websites** see the list on pages 8–9. These are all highly useful, specialist energy websites. Additionally, newspapers such as the *Independent*, *The Times*, the *Daily Telegraph* and the *Guardian* have excellent websites. Websites have the advantage of being up to date and to succeed in studying energy you need to be using topical materials. For example, the 2010 *Deepwater Horizon* disaster in the Gulf of Mexico has had a huge impact on the future of exploring 'new frontier' oil in deep water.

2 Summarise the articles you come across using case-study cards. File these either by energy source or by topic, using the headings in your specification. For each article, note the source and date.

3 When using different media, you have to consider the issue of **bias**. For many of the key topics in this book, such as the nuclear debate, there are a number of very different opinions, so you need to know the provenance of your information and data. Clearly British Nuclear Fuels will have a different opinion from anti-nuclear campaign groups. In the case of wind farms, the environmental groups cannot even agree whether wind is a green source of power (see page 81 **pros and cons of wind**). Sometimes there is different bias relating to scale, such as national versus local interest — for instance, 80% of people approve of wind farms, but not in their backyard (nimbyism).

The issue of the future of coal is particularly controversial, as the following headlines show:
- 'The great coal hole' — disputes the availability of coal reserves (*New Scientist*, January 2008).
- 'The big clean up' — supports the case for clean coal (*New Scientist*, September 2005).
- 'Brave new world for coal' — argues in favour of a future for clean coal (*Guardian*, January 2007).
- 'Old King Cole is a brave old soul but he is talking utter nonsense' — argues against coal on environmental grounds (*Guardian*, August 2008).
- 'A dirty business' — argues against coal on social grounds (*Guardian*, November 2005).

- 'Everything hinges on stopping coal' — an article from the climate camp strongly against coal (*Guardian*, August 2008).
- 'Coal's bright future' — in favour of coal as oil and gas supplies dwindle (*The Times*, July 2006).
- 'Cheap coal threat to global climate' — argues against use of coal for climate change reasons (*New Scientist*, March 2007).

Try to sort out the controversies suggested by the headlines as to whether coal should be a fuel of choice for the future. Read pages 42–48 to expand your ideas.

One way to assess bias is to summarise the article, summarise the opinion of the author and then, after you have done some further research, summarise your own opinions about the issue.

You can also construct a bias scale to show opinions for and against an issue.

Understanding energy units

In energy literature, you will come across a wide range of different units. Take care when interpreting graphs and check what each axis is showing. Some units refer to the weight (e.g. tonnes) or volume (e.g. m³) while others refer to the energy that can be released from energy sources (BTUs —British thermal units).

Some units, such as barrels and horsepower, are historical. Other older units include BTUs, which are still widely used in the UK and USA, and then the standard SI units widely used by the scientific community. Examples and definitions are listed below.

barrel (used for oil) = 158.98 litres

BTU (British thermal unit) = 1.06 kilojoules

btoe (billions of tonnes of oil equivalent)

kWh (kilowatt hour)

TW (terawatt) = 1 trillion watts

MJ (megajoule) = 1 million joules

GJ (gigajoule) = 1 billion joules

TJ (terajoule) = 1 trillion joules

m³ (cubic metres) — the usual measurement for gas volume

Note: In an exam question you may not recognise the units used on a graph but you will be able to access the graph enough to interpret trends and relationships. Do not be put off by unknown units.

Index

Note: page numbers in *italics* refer to information in tables and diagrams

A

accidents
  nuclear reactors *71*
  oil rigs *53*, 60
Africa
  energy poverty 11
  oil exploitation 66
  solar energy potential 102
  see also Nigeria; Sahara; South Africa
Alaska, oil exploitation 62
Amnesty International 67
Arctic, oil exploration 61–2
Asia
  nuclear power 77–8
  see also China; India; Japan
Asia-Pacific Partnership, energy mix 32–4
atmospheric pollution
  fossil fuels 38
  geothermal energy 88–9
Australia, energy mix *33*, 34

B

Bangladesh, solar power 85–6
Belarus, oil and gas disputes 18
Berkeley, CA, eco-housing 96
biofuels 26, 80, 83–5
biomass *14*, *80*
  biofuels from *see* biofuels
  cooking stoves 86
  in electricity generation 40, 41
  environmental impact *39*
  socioeconomic impact *39*
Brazil
  energy security 24
  offshore oil 60–1

buildings, energy efficiency 96–7
Bush, George 25

C

Cairn Oil 61–2
Canada, tar sands 64–5
carbon capture and sequestration 47–8
carbon dioxide emissions
  Asia-Pacific 32, *33*
  control and reduction 32, *95*, 100–1
  geothermal energy 89
  oil shale exploitation 65
  synthetic oil production 65
  tar sands exploitation 64
carbon footprint 40
  biofuels 84
  electricity generation 40–2, 71
carbon-neutral technologies 7, 40
  see also renewable energy
carbon pricing 7, 75
China
  carbon efficiency 100–1
  climate change 100–1
  coal 30, 42, 44, 45
  emissions reduction 100, 101
  energy insecurity 34
  energy mix 30, 33–4, 100
  environmental concerns 44
  gas 19
  HEP 83, 87
  nuclear power 72–3
  renewable energy *80*, 82, *100*
clean energy 38
  see also renewable energy
climate change 100–1

Contemporary Case Studies

G

gas *see* natural gas
Gas Exporting Countries Forum 20
gasification 47
gas industry, key players 57
gas pipelines 18, 19–20
Gazprom 18, 19, 20, 21, *57*
GDP 7
  and energy consumption 28, *29*
geo-engineering 7
geopolitics 16, 23, 94
  and gas supplies 18
  and oil supplies 21–2, 52–3
geothermal energy *14*
geothermal power *39*, *80*, 87–9
Germany, renewable energy *80*
governments, and oil industry *57*, 66
greenhouse gases *see* carbon dioxide emissions
Greenland, offshore oil 61–2
Greenpeace 62
green taxes 32, *95*
grid, electricty 7, 98–9
  see also off-grid communities

H

homeowners
  electricity generation 96–7
  smart meters 98, 99
housing, energy efficiency 96–7
human development index (HDI) 7
hydroelectric power 82–3
  availability *15*
  carbon footprint 41
  characteristics *14*
  environmental impact *39*, 83
  relative importance 31
  socioeconomic impact *39*

I

Iceland, geothermal power 87–9
imports, dependence on 25, 35, 36–7

India
  energy mix *33*, 34
  irrigation 87
  nuclear power 77
industrialisation, and energy sources 30
International Energy Agency 28, 54, 58
Iraq, oil industry 63
irrigation, solar-powered 87
Italy, electricity energy mix 72

J

Japan
  energy mix *33*, 34
  nuclear power 77–8

K

Korean Republic, oil supplies 24
Kyoto Protocol 15, 32, *95*

L

legislation, impact on energy mix 32
LNG (liquefied natural gas) 8
low-carbon technologies 40
  for electricity generation 41, 46–8
  see also nuclear power; renewable energy

M

market liberalisation 7, 101–2
mining
  coal *39*, 43, 45–6
  tar sands 64
  uranium *71*
modernisation theory 29–30

N

Nabucco pipeline 19–20
natural gas
  availability *15*
  characteristics *14*
  combustion products *38*

Contemporary Case Studies